Mathematical Modelling

Mathematical Modelling

A Graduate Textbook

Seyed M. Moghadas
York University, Canada

Majid Jaberi-Douraki
Kansas State University, USA

Registered Office(s)
John Wiley & Sons, Inc., 111 River Street, Hoboken, NJ 07030, USA

Editorial Office
111 River Street, Hoboken, NJ 07030, USA

For details of our global editorial offices, customer services, and more information about Wiley products visit us at www.wiley.com.

Library of Congress Cataloging-in-Publication Data
Names: Moghadas, Seyed M., author. | Jaberi-Douraki, Majid, author.
Title: Mathematical modelling : a graduate textbook / by Seyed M. Moghadas, Majid Jaberi-Douraki.
Description: 1st edition. | Hoboken, NJ : John Wiley & Sons, 2018. | Includes bibliographical
   references and index. |
Identifiers: LCCN 2018008846 (print) | LCCN 2018013540 (ebook) | ISBN 9781119484028 (pdf) |
   ISBN 9781119483991 (epub) | ISBN 9781119483953 (cloth)
Subjects: LCSH: Mathematical models–Textbooks.
Classification: LCC TA342 (ebook) | LCC TA342 .M636 2018 (print) | DDC 511/.8–dc23
LC record available at https://lccn.loc.gov/2018008846

Cover Design: Wiley
Cover Image: © Oli Kellett/Gettyimages

Set in 10/12pt Warnock by SPi Global, Pondicherry, India

Printed in the United States of America

V10002508_071318

To our parents

Contents

Preface

Mathematical modelling has evolved to become an important tool for understanding the underlying mechanisms of real-life problems, most notably in biological and medical sciences. With advances in computational methods and computing power, the last two decades have witnessed the emergence of a new generation of models that have opened up novel vistas for mathematics and statistics to play an ever more significant role in many disciplines, enhancing research efforts for new discoveries, from basic sciences to more applied and practical settings. Tracing back to its historical roots, the importance of mathematical modelling was eloquently characterized by Daniel Bernoulli (1760) in his work to predict the gain in life expectancy that would result from controlling smallpox, a baneful disease of humans at the time. This characterization was highlighted in his statement, "I simply wish that, in a matter which so closely concerns the well-being of mankind, no decision shall be made without all the knowledge which a little analysis and calculation can provide." More than 250 years later, this quote still remains a compelling argument for the use of mathematical models to generate new knowledge in various fields, including biology, physiology, ecology, finance, and medicine.

Due to the widening scope and application of mathematical models in various disciplines, there has been a surge of interest in training highly qualified personnel with professional skills and expertise in mathematical modelling. This training often starts with a course in this subject. The purpose of this book is to provide essential materials for an enriched training that is suitable for both undergraduate and graduate levels. While we strive to provide the necessary background and theoretical foundation for topics covered in this book, readers and researchers may consult the references for other advanced textbooks for more details on related topics. Our aim here is to demonstrate a wide spectrum of problems that can be addressed through mathematical modelling with the use of fundamental tools and techniques in applied mathematics and statistics. However, it should be emphasized that many of the theoretical concepts utilized throughout the book require sufficient knowledge of applied mathematics, and the audience would be advised to acquire the relevant materials from topic-specific textbooks, especially in differential equations and linear algebra.

This textbook contains nine chapters, starting with a chapter that provides the basic concepts of mathematical modelling, and a brief review of the relevant topics from differential equations and linear algebra. Chapters 2–5 elaborate on different types of mathematical models, and describe various techniques from dynamical systems theory for their analysis with a wide variety of examples. The analytical techniques are concerned with compartmental modelling, stability, bifurcation, discretization, and fixed-point analysis. The theoretical analyses in these chapters primarily involve systems of ordinary differential equations for deterministic models. Chapter 6 briefly reviews the concepts of probability and random variables, which are required for the stochastic processes and Markov chains presented in Chapter 7. In Chapter 8 we describe algorithms for computer simulations of both deterministic and stochastic models, and provide examples of Matlab© codes for simulating models. Finally, in Chapter 9, a number of well-known models are detailed to illustrate their application in different fields of study. Each chapter contains a number of exercises to help students better understand the modelling process, and develop enquiring and creative minds in this important subject.

This textbook is suitable for a modelling course in applied sciences at both undergraduate and graduate levels. It is also accessible to more theoretically oriented bioscientists who have some knowledge of linear algebra, systems of differential equations, and probability. In contextualizing this textbook, we have made a special effort to develop an interdisciplinary content for fundamentals and applications of mathematical modelling. The contents have been taught by the authors in one-semester mathematical modelling courses at the graduate level, and therefore benefited from continual improvements and feedback provided by students over a number of years. We hope that this textbook will be a resource that eliminates the need for a collection of multiple books by instructors and students in order to cover a number of important topics in a mathematical modelling course.

November 2017

Seyed M. Moghadas
York University, Canada

Majid Jaberi-Douraki
Kansas State University, USA

About the Companion Website

www.wiley.com/go/Moghadas/Mathematicalmodelling

The Companion Website provides the solution manual, presenting a guideline for solutions to the exercises within each chapter of the textbook. All solutions are based on the methods and techniques discussed in the textbook.

1

Basic Concepts and Quick Review

A standard scientific practice is to formulate an explanation for an observed phenomenon and then test this formulation by projecting the outcomes of various experiments under pertinent conditions. Projections are generally compared with experimental data. If there is agreement, the explanation can be accepted as a valid theory, whereas discrepancies point to a need for reformulation of the explanation. A model that describes the main features of the phenomenon, often represented mathematically, can be iteratively improved in the process of reformulation to resolve its discrepancies with observations or experimental data. This iterative process is known as the *modelling cycle* (Figure 1.1).

In simple terms, mathematical modelling is a process by which we derive a model to describe a phenomenon that may or may not be observable. For example, the movement of a pendulum is an observable phenomenon, but the transmission of a disease in the population may not be observable. In the latter case, the outcomes of infection and illness indicate that the epidemic phenomenon may be taking place and the disease is being transmitted among individuals. The process of modelling consists of several important steps. In general, the model represents a framework that includes simplification, assumptions, and approximation to describe the phenomenon under investigation. This framework can be expressed by mathematical equations and analyzed using the theory of dynamical systems and computational tools for model validation and comparison with available data (Figure 1.1).

Before proceeding further, let us present an example of developing a simple mathematical model. In this example, we wish to calculate the volume of sand that falls from the top half to the bottom half of a conical hourglass within a period of time (Figure 1.2). Suppose that the sand flows at the rate of 4 cm³ per second from the top half to bottom half of the hourglass. We remember from calculus that the volume of a cone with height h and radius R is given by $V = \pi R^2 h/3$. Here, we will first find the volume of sand in the bottom half of

Mathematical Modelling: A Graduate Textbook, First Edition. Seyed M. Moghadas and Majid Jaberi-Douraki.
© 2019 John Wiley & Sons, Inc. Published 2019 by John Wiley & Sons, Inc.
Companion Website: www.wiley.com/go/Moghadas/Mathematicalmodelling

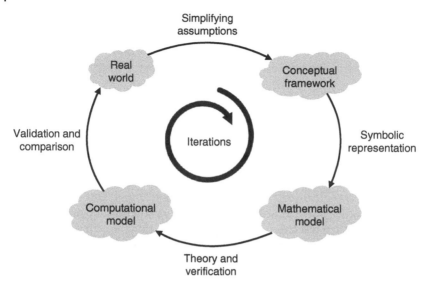

Figure 1.1 The process of model development, analysis, and validation.

the conical hourglass. From the dimensions given in Figure 1.2, this volume is given by:

$$V_h(t) = \frac{\pi 6^2 \times 12}{3} - \frac{\pi r^2 (12 - h)}{3} = 144\pi - \frac{\pi r^2 (12 - h)}{3}.$$

Using the property of similar triangles, we can write r in terms of h as $r = 6 - h/2$. Substituting this into V_h and taking the derivative of V_h with respect to t, we get:

$$\frac{\mathrm{d}V_h}{\mathrm{d}t} = \pi \left(6 - \frac{h}{2}\right)^2 \frac{\mathrm{d}h}{\mathrm{d}t},$$

where we consider h to be a function of time t. Given the flow rate of sand (i.e., $V'_h = 4$), the change in the height of sand in the bottom-half of the conical hourglass with respect to time is:

$$\frac{\mathrm{d}h}{\mathrm{d}t} = \frac{4}{\pi \left(6 - \frac{h}{2}\right)^2}.$$

Using separation of variables and integrating the height equation gives:

$$\int_{h(0)}^{h(t)} \left(6 - \frac{h}{2}\right)^2 \mathrm{d}h = \int_0^t \frac{4}{\pi} \, \mathrm{d}t.$$

Thus,

$$-(12 - h)^3 \Big|_{h(0)}^{h(t)} = \frac{48t}{\pi}.$$

Figure 1.2 Representation of a conical
hourglass.

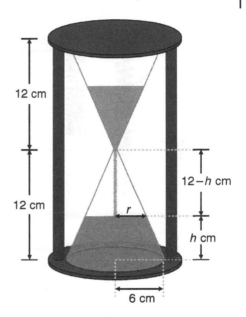

Since $h(0) = 0$, we have $h(t) = 12 - \sqrt[3]{12^3 - 48t/\pi}$. Substituting $h(t)$ into the
equation for $V_h(t)$, we can calculate the amount of sand that falls from the top
half to the bottom half of the conical hourglass within a certain time period.
For example, between $t = 0$ and $t = 36\pi$, the bottom half of the hourglass will
be filled, that is, $V(36\pi) = 144\pi$. This simple example shows how a model can
be used to describe the outcomes of a process that changes with time.

Mathematical models are often used to explore the dynamics of a system
over time. Let us present another example from classical mechanics. A simple
mechanical oscillating system can be illustrated by a weight attached to a linear
spring subject to only weight and tension, representing a harmonic oscillator. In
mechanics and physics, simple harmonic motion is a type of oscillation where
the restoring force is proportional to the displacement and acts in the direction
opposite to that of displacement. Ignoring the damping behavior, the restoring force (given by the product of mass and acceleration according to Newton's
second law of motion for a constant mass) in a linear spring can be modelled by:

$$F = ma = m\frac{d^2x}{dt^2} = mx'' = -kx, \tag{1.1}$$

where m is the mass attached to the spring, k is the spring constant, and x represents the displacement of the mass from its equilibrium state (Figure 1.3).
A solution of the equation $mx'' = -kx$ (with the initial condition $x(0) = 0$) is
given by:

$$x(t) = A \sin\left(\sqrt{\frac{k}{m}}\, t\right),$$

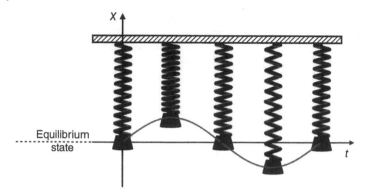

Figure 1.3 Representation of spring motion in a sinusoidal form.

where A is a constant representing the amplitude of sinusoidal motion of the spring. The displacement x is represented as a function of t in Figure 1.3.

1.1 Modelling Types

Broadly speaking, mathematical models can be classified based on three major characteristics that may depend on the nature of the phenomenon: (i) deterministic or stochastic; (ii) dynamic or static; and (iii) discrete or continuous (Figure 1.4). To better understand this classification, we provide some specific examples as follows.

- Deterministic of dynamic-continuous type: steam engine.
- Deterministic of static-continuous type: snapshot of pendulum.
- Deterministic of dynamic-discrete type: the percentage of computer processing unit in use upon startup.
- Deterministic of static-discrete type: clock cycles for a computer program to run on a given input.
- Stochastic of dynamic-continuous type: weather.
- Stochastic of static-continuous type: noise in an electronic circuit.
- Stochastic of dynamic-discrete type: random arrivals.
- Stochastic of static-discrete type: flipping a coin.

In this textbook, we present various examples of mathematical models within this classification and analyze their behavior. In our analyses of such models, we use essential analytical tools from the theory of dynamical systems. Here, we briefly review some techniques from the theory of differential equations and linear algebra that are useful in understanding the analytical tools and their applications to mathematical modelling in subsequent chapters.

Figure 1.4 Types of deterministic and stochastic models.

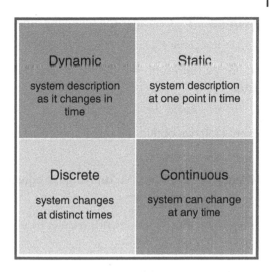

1.2   Quick Review

We begin by reviewing methods for solving first- and second-order linear differential equations [7].

1.2.1   First-order Differential Equations

The general form of a first-order linear differential equation is given by:

$$\frac{dy}{dx} + P(x)y = Q(x),\tag{1.2}$$

where $P(x)$ and $Q(x)$ are continuous real-valued functions. To solve this equation for a solution in the form of $y(x)$, we use the integration factor:

$$R(x) = \exp\left(\int_a^x P(t)dt\right).\tag{1.3}$$

Multiplying (1.2) by $R(x)$ gives:

$$\frac{d(R(x)y(x))}{dx} = R(x)Q(x).$$

Integrating both sides of this equation with respect to x gives:

$$R(x)y(x) = \int R(t)Q(t)dt + C,\tag{1.4}$$

where C is a constant. Since $R(x) \neq 0$ for all $x \in \mathbb{R}$, we can divide each side of (1.4) by $R(x)$ to obtain the solution of $y(x)$. To illustrate this method, we provide the following example.

Example 1.1 Consider the following differential equation:

$$y' = 2x(y + 1),$$

where we represent $\frac{dy}{dx}$ by y'. Rewriting this equation in general form of (1.2) gives:

$$y' - 2xy = 2x.$$

From (1.3), we obtain the integration factor:

$$R(x) = Ae^{-x^2},$$

where A is a constant. Multiplying the equation by $R(x)$ gives:

$$(e^{-x^2}y)' = 2xe^{-x^2}.$$

Integrating this equation with respect to x leads to the solution:

$$y = -1 + Ce^{x^2},$$

where C is a constant.

☞ **Bernoulli equation.** The general form of a first-order differential equation of Bernoulli type is given by [7]:

$$\frac{dy}{dx} + P(x)y = Q(x)y^n, \tag{1.5}$$

where $P(x)$ and $Q(x)$ are continuous real-valued functions, and $n = 0, 1, 2, \ldots$. If $n > 1$, then we can use the change of variable $u = y^{1-n}$. Thus, equation (1.5) reduces to:

$$\begin{aligned}
\frac{du}{dx} &= (1-n)y^{-n}\frac{dy}{dx} \\
&= (1-n)y^{-n}(-P(x)y + Q(x)y^n) \\
&= (1-n)(-P(x)y^{1-n} + Q(x)) \\
&= (1-n)(-P(x)u + Q(x)).
\end{aligned} \tag{1.6}$$

Equation (1.6) can now be solved using an integration factor.

Example 1.2 Consider the following differential equation:

$$y' - \frac{y}{x} = y^2 \ln x. \tag{1.7}$$

Letting $u = y^{-1}$, we get:

$$u' + \frac{u}{x} = -\ln x.$$

Using the integration factor $R(x) = x$, we get $(xu)' = -x \ln x$, and therefore:

$$u = -\frac{x}{2}(\ln x - \frac{1}{2}) + \frac{C}{x},$$

where C is a constant. Thus, the solution of (1.7) is:

$$y = \frac{4x}{-x^2(2\ln x + 1) + C}.$$

1.2.2 Second-order Differential Equations

The general form of a second-order linear differential equation is given by:

$$\frac{d^2y}{dx^2} + a(x)\frac{dy}{dx} + b(x)y = F(x), \tag{1.8}$$

where $F(x)$, $a(x)$ and $b(x)$ are continuous real-valued functions. Here, we assume that $a(x) = a$ and $b(x) = b$ are constants. We can consider homogeneous and inhomogeneous cases.

☞ **Homogeneous case.** In this case, $F(x) \equiv 0$, and (1.8) reduces to:

$$\frac{d^2y}{dx^2} + a\frac{dy}{dx} + by = 0. \tag{1.9}$$

We look for a solution of the form $y(x) = e^{\lambda x}$, in which λ is yet to be determined. Substituting this solution into (1.9) gives the characteristic equation:

$$(\lambda^2 + a\lambda + b)e^{\lambda t} = 0.$$

Solving this equation for λ will provide different types of solutions for (1.9), depending on whether the characteristic equation has distinct real roots, repeated roots, or complex roots. The general form of the solution of (1.9) for distinct roots is then given by:

$$y(x) = C_1 e^{\lambda_1 x} + C_2 e^{\lambda_2 x},$$

and for repeated roots by:

$$y(x) = C_1 e^{\lambda x} + C_2 x e^{\lambda x}.$$

Example 1.3 Consider the following second-order differential equation:

$$y'' - y' - 2y = 0.$$

Solving the characteristic equation $\lambda^2 - \lambda - 2 = 0$ gives the solutions $\lambda_1 = -1$, and $\lambda_2 = 2$. Therefore, the general solution of (1.9) can be expressed by:

$$y(x) = C_1 e^{-x} + C_2 e^{2x},$$

where C_1 and C_2 are constants.

Example 1.4 Consider the following second-order differential equation:

$$y'' - 2y' + y = 0.$$

Solving the characteristic equation $\lambda^2 - 2\lambda + 1 = 0$ gives the solutions $\lambda_1 = \lambda_2 = 1$. Therefore, the general solution of (1.9) can be expressed by:

$$y(x) = C_1 e^x + C_2 x e^x,$$

where C_1 and C_2 are constants.

☞ **Inhomogeneous case.** In this case, $F(x) \neq 0$. We first solve the homogeneous equation (setting $F(x) = 0$), and then extend this solution to the inhomogeneous case. Suppose $y_h(x)$ represents the solution for the homogeneous equation. We consider a particular solution $y_p(x)$ in a similar functional form to $F(x)$ with unknown constants. The general form of the solution for inhomogeneous case is then given by:

$$y(x) = y_h(x) + y_p(x).$$

In the last step, we find the coefficients of the particular solution by substituting $y_p(x)$ into equation (1.8).

Example 1.5 Consider the following second-order differential equation:

$$y'' - y' - 2y = x^2. \tag{1.10}$$

Solving the homogeneous case $y'' - y' - 2y = 0$ as described above gives the solution:

$$y_h(x) = C_1 e^{-x} + C_2 e^{2x}.$$

We now assume that the particular solution has the form of a polynomial of degree 2 similar to the functional form of $F(x) = x^2$:

$$y_p(x) = Ax^2 + Bx + C.$$

Substituting this particular solution into equation (1.10), we get:

$$2A - (2Ax + B) - 2Ax^2 - 2Bx - 2C = x^2.$$

Rearranging this equation, we find that the constant term and the coefficient of x must be zero, and the coefficient of x^2 must be 1, so that the equation holds for all $x \in \mathbb{R}$. This implies that:

$$-2A = 1,$$
$$-2B - 2A = 0,$$
$$2A - B - 2C = 0.$$

Thus, $A = -\frac{1}{2}$, $B = \frac{1}{2}$, and $C = -\frac{3}{4}$. Hence, we obtain the general solution of (1.10):

$$y(x) = C_1 e^{-x} + C_2 e^{2x} - \frac{1}{2}x^2 + \frac{1}{2}x - \frac{3}{4}.$$

Example 1.6 Consider the following differential equation:

$$y'' - y = 2e^{5x}. \tag{1.11}$$

Solving the characteristic equation for the homogeneous case gives the solution $y_h(x) = C_1 e^{-x} + C_2 e^x$. Assuming $y_p(x) = Ae^{5x} + B$, and substituting back into (1.11), we find $A = \frac{1}{12}$ and $B = 0$. Thus, the general solution of (1.11) is:

$$y(x) = C_1 e^{-x} + C_2 e^x + \frac{1}{12}e^{5x}.$$

Example 1.7 Consider the following inhomogeneous differential equation:

$$y'' + y' + y = x + \cos x. \tag{1.12}$$

The characteristic equation for the homogeneous case is $\lambda^2 + \lambda + 1 = 0$, which has the solutions $\lambda_\pm = (-1 \pm i\sqrt{3})/2$. This gives the solutions in the complex domain as $y(x) = e^{-\frac{x}{2}} \left(C_1 e^{i\frac{\sqrt{3}x}{2}} + C_2 e^{-i\frac{\sqrt{3}x}{2}} \right)$. Therefore, we find the solution:

$$y_h(x) = C_1 e^{-\frac{x}{2}} \sin \frac{\sqrt{3}}{2}x + C_2 e^{-\frac{x}{2}} \cos \frac{\sqrt{3}}{2}x.$$

We now consider a particular solution of the form $y_p(x) = Ax + B + D\sin x + E\cos x$. Substituting $y_p(x)$ into (1.12) gives:

$$-E\sin x + D\cos x + Ax + A + B = x + \cos x,$$

which implies:

$$A = 1, \quad B = -1, \quad E = 0, \quad D = 1.$$

Thus, the general solution of the equation is:

$$y(x) = C_1 e^{-\frac{x}{2}} \sin \frac{\sqrt{3}}{2}x + C_2 e^{-\frac{x}{2}} \cos \frac{\sqrt{3}}{2}x + x - 1 + \sin x.$$

1.2.3 Linear Algebra

A number of analytical tools that we introduce in the subsequent chapters apply fundamental concepts from matrix theory and linear algebra [49]. Here, we provide an overview of these concepts.

A square matrix $A_{n\times n}$ (with equal number of rows and columns) is invertible, if there exists a square matrix $B_{n\times n}$ such that $AB = BA = I_{n\times n}$, where

$$I_{n\times n} = \begin{bmatrix} 1 & 0 & 0 & \cdots & 0 \\ 0 & 1 & 0 & \cdots & 0 \\ \vdots & \vdots & \vdots & \ddots & \vdots \\ 0 & 0 & 0 & \cdots & 1 \end{bmatrix}_{n\times n}.$$

The matrix B is often denoted by A^{-1} (referred to as the inverse of A), and it is unique. It is important to note that the concept of inverse applies only to square matrices, while it is possible to find matrices A and B that satisfy the condition of $AB = I$ but are not square. For example, let:

$$A = \begin{bmatrix} 1 & 0 & 2 \\ 0 & 0 & 2 \end{bmatrix}, \qquad B = \begin{bmatrix} 1 & -1 \\ 0 & 0 \\ 0 & \frac{1}{2} \end{bmatrix}.$$

Then multiplication of A and B gives:

$$AB = \begin{bmatrix} 1 & 0 \\ 0 & 1 \end{bmatrix} = I_{2\times 2}.$$

However, neither A nor B is invertible because:

$$BA = \begin{bmatrix} 1 & 0 & 0 \\ 0 & 0 & 0 \\ 0 & 0 & 1 \end{bmatrix} \neq AB.$$

For a square matrix $A_{n\times n}$, there is a number called the *determinant*, denoted by $\det(A)$. According to the matrix theory, a square matrix is invertible (i.e., A^{-1} exists) if and only if $\det(A) \neq 0$. Using the inverse of a matrix, when it exists, it is possible to solve systems of linear equations. To illustrate this, let us consider the general form of a system of n linear equations:

$$a_{11}x_1 + a_{12}x_2 + \cdots + a_{1n}x_n = b_1,$$
$$\vdots \qquad\qquad \vdots \qquad\qquad (1.13)$$
$$a_{n1}x_1 + a_{n2}x_2 + \cdots + a_{nn}x_n = b_n.$$

We can rewrite (1.13) in matrix form as $AX = B$, where

$$A = \begin{bmatrix} a_{11} & \cdots & a_{1n} \\ \vdots & \ddots & \vdots \\ a_{n1} & \cdots & a_{nn} \end{bmatrix}, \quad X = \begin{bmatrix} x_1 \\ \vdots \\ x_n \end{bmatrix}, \quad B = \begin{bmatrix} b_1 \\ \vdots \\ b_n \end{bmatrix}.$$

If $\det(A) \neq 0$, then A^{-1} exists, and we can multiply each side of the equation by A^{-1} to get:

$$A^{-1}(AX) = A^{-1}B.$$

Using the associative property of matrix multiplication, and $AA^{-1} = I$, we obtain the solution of the system as $X = A^{-1}B$.

For a square matrix $A_{n \times n}$, a scalar λ is called an eigenvalue if there is a nonzero solution X of $AX = \lambda X$. Such an X is called an *eigenvector* corresponding to the *eigenvalue* λ. The eigenvalues and the corresponding eigenvectors are obtained by solving the characteristic equation:

$$\det(A - \lambda I) = 0,$$

where λ is an eigenvalue of matrix A. For each λ, a vector $V \neq 0$ is an eigenvector if:

$$(A - \lambda I)V = 0.$$

Example 1.8 Find the eigenvalues and the corresponding eigenvectors of the matrix:

$$A = \begin{bmatrix} 1 & 4 \\ 2 & -1 \end{bmatrix}.$$

For this purpose, we solve the characteristic equation:

$$\det \left(\begin{bmatrix} 1 & 4 \\ 2 & -1 \end{bmatrix} - \lambda \begin{bmatrix} 1 & 0 \\ 0 & 1 \end{bmatrix} \right) = \det \begin{bmatrix} 1 - \lambda & 4 \\ 2 & -1 - \lambda \end{bmatrix} = 0,$$

which is $-(1 + \lambda)(1 - \lambda) - 8 = 0$. Thus, there are two eigenvalues $\lambda_1 = 3$ and $\lambda_2 = -3$. To find an eigenvector V_1 corresponding to λ_1, we consider the equation $(A - 3I)V_1 = 0$, which gives the linear system:

$$\begin{bmatrix} -2 & 4 \\ 2 & -4 \end{bmatrix} \begin{bmatrix} x_1 \\ x_2 \end{bmatrix} = \begin{bmatrix} 0 \\ 0 \end{bmatrix}.$$

This system provides only one equation with two unknowns given by $-2x_1 + 4x_2 = 0$. Since we are looking for a nonzero vector, assuming $x_1 = 1$, we find $x_2 = 1/2$, and therefore:

$$V_1 = \begin{bmatrix} 1 \\ \frac{1}{2} \end{bmatrix}.$$

In a similar way, we can find an eigenvector corresponding to the eigenvalue $\lambda_2 = -3$. In this case, we need to solve the linear system $(A + 3I)V_2 = 0$. This system provides only one equation, given by $2x_1 + 2x_2 = 0$. Assuming $x_1 = 1$, we obtain $x_2 = -1$, and therefore a nonzero eigenvector is obtained as:

$$V_2 = \begin{bmatrix} 1 \\ -1 \end{bmatrix}.$$

1.2.4 Scaling

Scaling is a useful technique in mathematical modelling to simplify the model for analysis, without changing the theoretical structure of the system or its behavior. Scaling is often used to make independent and dependent variables dimensionless, normalize the size associated with the system variables, and reduce the number of independent parameters in the model [25]. Here, we explain this technique using the following examples.

Example 1.9 Consider the following differential equation:

$$\frac{dN}{dt} = r N \left(1 - \frac{N}{K}\right),$$

where r and K are positive numbers, and $N \geq 0$. This equation is known as the *logistic model*, and we will detail its properties in the next chapter. The variable N represents the size of a population which changes with time, and depends on the growth rate r and the carrying capacity K. We may simplify this equation by scaling through a new variable and a new parameter, defined by:

$$n = \frac{N}{K}, \quad \tau = rt.$$

Taking the derivative of n with respect to τ and using the chain rule, we get:

$$\frac{dn}{d\tau} = \frac{dn}{dt} \cdot \frac{dt}{d\tau}$$
$$= \frac{d}{dt}\left(\frac{N}{K}\right) \cdot \left(\frac{1}{r}\right) = \frac{1}{rK}\frac{dN}{dt}$$
$$= \frac{1}{rK}(rN)\left(1 - \frac{N}{K}\right)$$
$$= \frac{N}{K}\left(1 - \frac{N}{K}\right) = n(1 - n).$$

Thus, the logistic equation can be simplified to the equation $n' = n(1 - n)$, where $n \geq 0$.

Example 1.10 Consider the following system of nonlinear differential equations:

$$\frac{dS}{dt} = -\beta S(t)I(t),$$

$$\frac{dI}{dt} = \beta S(t)I - \gamma I(t).$$

This system may represent the spread of a disease in a population of susceptible individuals (S), with a transmission rate of β through contacts with infected individuals (I). Infected individuals recover at a rate of γ. We will delineate epidemic models in the following chapters. For this *SI epidemic model*, we set $S_0 = S(0)$, and define:

$$u = \frac{S(t)}{S_0}, \quad v = \frac{I(t)}{S_0}, \quad \tau = \gamma t.$$

Differentiating u with respect to τ gives:

$$\frac{du}{d\tau} = \frac{du}{dt} \cdot \frac{dt}{d\tau}$$

$$= \frac{d}{dt}\left(\frac{S}{S_0}\right) \cdot \frac{1}{\gamma} = \frac{1}{S_0\gamma}(-\beta SI)$$

$$= \frac{-\beta}{S_0\gamma}(uS_0)(vS_0)$$

$$= \frac{-\beta S_0}{\gamma}uv.$$

Similarly, we have:

$$\frac{dv}{d\tau} = \frac{dv}{dt} \cdot \frac{dt}{d\tau}$$

$$= \frac{d}{dt}\left(\frac{I}{S_0}\right) \cdot \frac{1}{\gamma}$$

$$= \frac{1}{S_0\gamma}[\beta SI - \gamma I]$$

$$= \frac{1}{S_0\gamma}[\beta(uS_0)(vS_0) - \gamma vS_0]$$

$$= \frac{\beta S_0}{\gamma}uv - v = \left(\frac{\beta S_0 u}{\gamma} - 1\right)v.$$

If we define $R_0 = \frac{\beta S_0}{\gamma}$, then the SI epidemic model reduces to:

$$\frac{du}{d\tau} = -R_0 uv,$$

$$\frac{dv}{d\tau} = (R_0 u - 1)v.$$

This simplified model, which depends on a single parameter R_0, could help us understand the behavior of the epidemic dynamics at the early stages of disease onset (i.e., for $t > 0$ and sufficiently small). We note that at the early stage of an epidemic, the number of infected individuals is small compared to the size of the susceptible population. Thus, it is reasonable to assume that for small t, $S(t) \approx S_0$ or $u \approx 1$. This assumption can be used to solve the equation $v' = (R_0 - 1)v$, representing the dynamics of the infected population at the early stages of the epidemic. Solving this equation, with an initial value of $v_0 = I(0)/S(0)$ gives:

$$\ln v(\tau) - \ln v_0 = (R_0 - 1)\tau.$$

Thus,

$$v(\tau) = v_0 e^{(R_0 - 1)\tau}.$$

This solution suggests that $v(\tau)$ decreases if $R_0 < 1$, and increases exponentially if $R_0 > 1$.

Exercises

1 Solve the following differential equations:
  a) $y' = \mu y + x^5$, where $\mu \in \mathbb{R}$ is a constant.
  b) $y'' - y = 2x + e^x$.
  c) $y'' - y = x + \sin x$.

2 Solve the following differential equation for distinct roots if $\alpha < 1$, repeated roots if $\alpha = 1$, and complex roots if $\alpha > 1$:

$$2y'' - 2\sqrt{2}y' + \alpha y = -\frac{1}{ax^2} - \frac{\sqrt{2}}{ax} + \frac{1}{2}\ln x + 5,$$

for $x > 0$. *Hint*: define $y_p(x) = A \ln x + B$.

3 Find the values of a for which the following matrix is invertible:

$$A = \begin{bmatrix} 0 & 1 & a \\ 1 & 3 & 0 \\ 0 & 2 & 2 \end{bmatrix}.$$

4 Find the eigenvalues and the corresponding eigenvectors of the matrix:

$$A = \begin{bmatrix} 0 & 1 & 1 \\ 0 & 3 & 1 \\ 1 & 0 & 2 \end{bmatrix}.$$

5 Solve the linear system:

$$\begin{bmatrix} 0 & 1 & 1 \\ 0 & 3 & 1 \\ 1 & 1 & 2 \end{bmatrix} \begin{bmatrix} x_1 \\ x_2 \\ x_3 \end{bmatrix} = \begin{bmatrix} 0 \\ \frac{1}{2} \\ 3 \end{bmatrix}.$$

6 Consider the following differential equation:

$$mx''(t) + kx(t) = 0, \quad x(0) = a, \quad x'(0) = b,$$

where m, k, a, and b are constant numbers. Use an appropriate change of variables to scale this equation into an equation with dimensionless independent and dependent variables.

2

Compartmental Modelling

Compartmental modelling is the most frequently employed approach in the study of system dynamics in several areas, including biology, epidemiology, pharmacokinetics, physiology, and chemical reactions. In this type of modelling, the structure of the system is simplified while attempting to preserve fundamental properties of the underlying dynamical processes. A compartmental model divides the system into a finite number of interconnected states to describe the system dynamics based on the net rate of change in the processes within each compartment. In this chapter, we describe compartmental modelling and provide examples from various disciplines.

☞ **Balance law.** According to this law, the net rate of change in a process is the difference between the input and output rates (Figure 2.1). This law plays a key role in modelling real-life and physical phenomena, and is expressed as:

{net rate of change in the process} = {rate of input} − {rate of output}.

Figure 2.1 Illustration of the balance law in modelling a process.

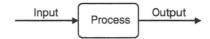

Example 2.1 It is well known that radioactive elements emit particles while they are decaying into isotopes of other materials over time (Figure 2.2). Suppose $N(t)$ is the number of radioactive nuclei at time t. Let Δt be a small change in time and k be the decay rate of nuclei per unit time. For a given number of radioactive nuclei, the process of decay involves only emissions (as output) over time. Thus, from the balance law, the net rate of change in $N(t)$ over time can be written as:

$$N(t + \Delta t) - N(t) = -kN(t)\Delta t. \tag{2.1}$$

Mathematical Modelling: A Graduate Textbook, First Edition. Seyed M. Moghadas and Majid Jaberi-Douraki.
© 2019 John Wiley & Sons, Inc. Published 2019 by John Wiley & Sons, Inc.
Companion Website: www.wiley.com/go/Moghadas/Mathematicalmodelling

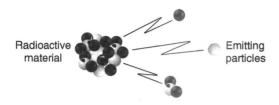

Figure 2.2 Illustration of radioactive material emitting particles.

Radioactive material

Emitting particles

Rewriting (2.1) in the form:

$$\frac{N(t + \Delta t) - N(t)}{\Delta t} = -kN(t),$$

and taking the limit when Δt approaches 0, we obtain the following differential equation:

$$\frac{dN}{dt} = -kN(t). \tag{2.2}$$

Let $N(t_0) = n_0$ be the number of radioactive nuclei at time t_0. Integrating (2.2) from t_0 to $t > t_0$ gives $\ln(N(t)) - \ln(n_0) = -k(t - t_0) + C$, where C is a constant. Using the condition at t_0, it follows that $e^C = 1$, and therefore $N(t) = n_0 e^{-k(t-t_0)}$.

☞ **Half-life.** Half-life is defined as the length of time required for half of the nuclei to decay. Using the model for radioactive nuclei, one can determine the half-life based on the decay rate of the material. Mathematically, we need to find τ, where $N(t) = 2N(t + \tau)$, and therefore:

$$\frac{n_0 e^{-k(t+\tau-t_0)}}{n_0 e^{-k(t-t_0)}} = \frac{1}{2}.$$

This implies that $\tau = \ln 2/k$. We note that the half-life (τ) is independent of the number of radioactive nuclei at any point in time.

In compartmental modelling, the rates associated with the system dynamics often correspond to the length of time required for a specific process to complete. For instance, in the example of radioactive material emitting isotopes, we considered a rate of decay. To better understand the meaning of this rate, we note that $\frac{N(t+t_0)}{n_0} = e^{-kt}$ represents the fraction of particles that are still in the compartment at time $t + t_0$. Let us view this fraction as the probability of being in the compartment of radioactive material at time t. Thus, the cumulative probability that particles have left the compartment at time t is given by $F(t) = 1 - e^{-kt}$. The probability density function (which we will learn more about in Chapter 6), $f(t)$, is defined as the derivative of the cumulative probability $F(t)$, and is given by:

$$f(t) = F'(t) = ke^{-kt}.$$

This shows that $f(t)$ has a positive exponential form and its mean over time is given by:

$$\int_0^\infty tf(t) = \int_0^\infty tke^{-kt} = -\left[te^{-kt} + \frac{e^{-kt}}{k}\right]\Big|_0^\infty = \frac{1}{k},$$

and thus $1/k$ describes the mean time that an individual particle spends in the compartment.

Example 2.2 In physiology and pharmacokinetics, compartmental modelling can be used to understand the effect of drug treatment. A simple example of such modelling relates to drug absorption by the gastrointestinal (GI) tract and diffusion into the bloodstream. Once diffused, the drug is transferred to the location targeted for treatment and is removed from the blood by the filtering system of the kidneys. A compartmental representation of these processes is shown in Figure 2.3.

To formulate the model, let $x(t)$ be the amount of drug absorbed and diffused into the bloodstream at time t. Assuming I is the initial drug intake, and κ is the rate of diffusion of drug, the net rate of change in the GI tract compartment is expressed by the following differential equation:

$$\frac{dx}{dt} = I - \kappa x, \quad x(0) = 0. \tag{2.3}$$

Suppose $y(t)$ is the concentration of drug in the bloodstream at time t, which is reduced by filtering at a rate r. The dynamics of drug concentration is governed by the following differential equation:

$$\frac{dy}{dt} = \kappa x - ry, \quad y(0) = 0. \tag{2.4}$$

Using the integration factor $e^{\kappa t}$ and solving the linear equation (2.3), it follows that:

$$x(t) = \frac{I}{\kappa}(1 - e^{-\kappa t}).$$

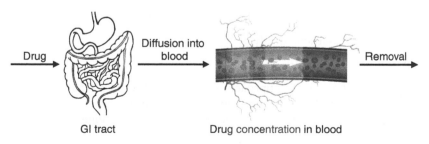

Drug — GI tract — Diffusion into blood — Drug concentration in blood — Removal

Figure 2.3 Compartmental representation of drug absorption by the GI tract and diffusion into the bloodstream.

Substituting $x(t)$ into (2.4) and similarly using the integration factor e^{rt}, we obtain:

$$\int_0^t (ye^{rs})' ds = \int_0^t I(1 - e^{-\kappa s}) e^{rs} ds,$$

which gives the concentration of drug in the bloodstream at time t as:

$$y(t) = \frac{I}{r}\left[1 - \frac{1}{r - \kappa}(re^{-\kappa t} - \kappa e^{-rt})\right].$$

2.1 Cascades of Compartments

Compartmental modelling can be classified into two types: linear and branching cascades (Figure 2.4). We provide some examples to discuss these cascades.

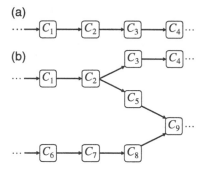

(a)

(b)

Figure 2.4 Linear (a) and branching (b) types of cascades in compartmental modelling.

Example 2.3 (Linear cascade in population growth) Suppose $N(t)$ describes the population size at time t. Let b represent the rate of birth (or immigration) and d be the rate of death (or emigration). The rate of change in the population size with the linear cascade shown in Figure 2.5 is:

$$\{\text{change in population size}\} = \{\text{birth or immigration}\}$$
$$-\{\text{death or emigration}\},$$

which is expressed mathematically by:

$$N(t + \Delta t) - N(t) = (b - d)N(t)\Delta t.$$

Taking the limit when Δt approaches 0, we get $N'(t) = rN(t)$, where we assume $r = b - d > 0$. With this equation, the population size at any point in time is $N(t) = N_0 e^{rt}$, where $N_0 = N(0)$, representing an exponential growth (Figure 2.6).

☞ **Logistic equation.** Realistically, populations cannot grow unboundedly in size (as predicted by the exponential growth) due to a number of factors such

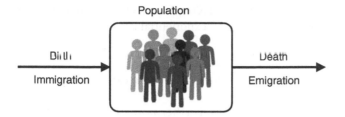

Figure 2.5 Linear cascade for compartmental modelling of population growth.

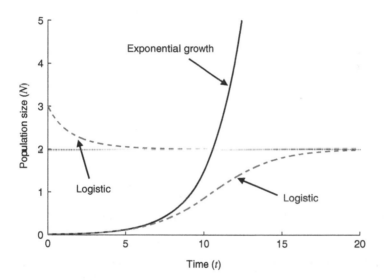

Figure 2.6 Population size over time in exponential growth and logistic models with $r = 0.5$ and $K = 2$.

as availability of resources. Therefore, a limiting population size may be considered, even when the population initially grows exponentially. This limiting value is called the *carrying capacity*. The logistic equation, initially developed by Pierre-François Verhulst in 1838, proposes a model to include this carrying capacity in the population growth model [53]. In the logistic equation, for a small number of individuals, the growth behaves exponentially, while for a large number of individuals, the population saturates at its carrying capacity.

Subtracting the term $cN^2(t)$, with $c > 0$ constant, from the right-hand side of the model with exponential growth gives $N'(t) = rN(t) - cN^2(t)$. Defining the carrying capacity by $K = r/c$, the logistic equation is obtained as:

$$N' = rN \left(1 - \frac{N}{K}\right). \tag{2.5}$$

When N is small compared to K, the term $N(t)/K$ is negligible and therefore population size can be estimated using the exponential growth model. However, when N is large, we consider two possibilities:

1) If $N < K$, then $N' > 0$ and $\lim_{t \to \infty} N'(t) = 0$. This shows that $N(t)$ saturates at a constant value. This value, as we show later, is the carrying capacity K (Figure 2.6).
2) If $N > K$, then $N' < 0$ and N decreases. Furthermore, $\lim_{t \to \infty} N' = 0$ and, similarly to the case of $N < K$, the population size saturates at its carrying capacity K (Figure 2.6).

To solve the logistic equation, we rewrite (2.5) in the form:

$$\left(\frac{1}{N} + \frac{1}{K - N} \right) dN = r dt.$$

Integrating each side from 0 to t, we obtain:

$$\ln \left(\frac{N(t)}{K - N(t)} \right) \Bigg|_0^t = rt + A,$$

where A is a constant. Assuming $N(0) = N_0$, we get:

$$N(t) = \frac{K N_0 e^{rt}}{K + N_0 (e^{rt} - 1)}. \tag{2.6}$$

Taking the limit of (2.6) when $t \to \infty$ gives $\lim_{t \to \infty} N(t) = K$.

Example 2.4 (Linear cascade in epidemic dynamics)   Suppose a communicable disease is introduced into a population through some initial number of infections [8]. The infection will spread through contacts between susceptible and infected individuals. We assume that the disease can be contained through a recovery process, leading to permanent immunity of individuals against re-infection. This simple epidemic dynamics can be shown by a linear cascade of compartments describing the population size of susceptible individuals (S), infected individuals (I), and recovered individuals (R) at time t (Figure 2.7).

We define β as the rate of disease transmission and γ as the rate of recovery from infection. The dynamics of infection is then governed by the following system of nonlinear differential equations:

$$S' = \underbrace{-\beta S I}_{\text{occurrence of infection}}$$

$$I' = \underbrace{\beta S I}_{\text{new infections (incidence)}} - \underbrace{\gamma I}_{\text{recovery from infection}} \tag{2.7}$$

$$R' = \underbrace{\gamma I}_{\text{recovered (immune)}}$$

Figure 2.7 Linear cascade of compartmental modelling for the dynamics of disease spread in a population.

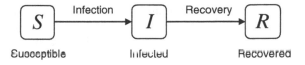

The solutions of this system cannot be expressed in closed form. However, we can explore some fundamental properties of the system. Suppose $S(0) = S_0$, $I(0) = I_0$ and $R(0) = 0$. As shown in Section 1.2.4, we can use the change of variables $u = S/S_0$, $v = I/S_0$, $w = R/S_0$, and $\tau = \gamma t$ to scale the system (2.7), which gives:

$$u' = \frac{-\beta S_0}{\gamma} uv,$$

$$v' = \left(\frac{\beta S_0 u}{\gamma} - 1 \right) v,$$

(2.8)

$$w' = v.$$

Similar to Example 1.10 in Chapter 1, we assume that $\frac{S(t)}{S_0} \approx 1$ at the early stages of the epidemic (i.e., $t > 0$ sufficiently small). Thus, defining $R_0 = \beta S_0 / \gamma$, the number of infections at time t (referred to as the *prevalence* of disease) can be estimated by:

$$v(t) = v(0)e^{(R_0-1)t}.$$

We should point out that this equation is valid only for small t, while $\frac{S(t)}{S_0} \approx 1$ holds. This indicates that when $R_0 < 1$, the number of infected individuals is expected to decline initially, while for $R_0 > 1$, the number of infections grows exponentially. We now define an important epidemiological concept that is related to the parameter R_0.

Definition 2.1 The *basic reproduction number* is defined as the average number of new infections generated by a single infected individual introduced into an entirely susceptible population [6]. This quantity is commonly denoted by R_0. If $R_0 < 1$, then no new infection will occur and the epidemic dies out. If $R_0 > 1$, then new infections will occur and the epidemic is expected to unfold.

An important question arises when the epidemic unfolds (i.e., $R_0 > 1$): how many individuals will be infected throughout the epidemic?

The total number of infected individuals during the epidemic is called the *final size*. We will find the final size by calculating the number of individuals who are still susceptible at the end of the epidemic. Theoretically, the end of the epidemic is defined when t is sufficiently large (i.e., $t \to \infty$).

Let $u_\infty = \lim_{t\to\infty} u(t)$ and $v_\infty = \lim_{t\to\infty} v(t)$. Adding the first and second equations in (2.8) gives $(u + v)' = -v$. By integrating this equation from 0 to ∞, and assuming that at the end of the epidemic $v_\infty = 0$, we get:

$$\int_0^\infty (u + v)' dt = - \int_0^\infty v(t)dt,$$

and therefore,

$$u_\infty - u_0 - v_0 = - \int_0^\infty v(t)dt. \tag{2.9}$$

From the first equation in (2.8), we have $\frac{u'}{u} = -R_0 v$. Integrating this equation from 0 to ∞ gives:

$$\ln\left(\frac{u_\infty}{u_0}\right) = -R_0 \int_0^\infty v(t)dt. \tag{2.10}$$

Comparing (2.9) and (2.10), we arrive at the *final size relation* for the epidemic model (2.7):

$$\ln\left(\frac{u_\infty}{u_0}\right) = R_0(u_\infty - u_0 - v_0). \tag{2.11}$$

For given initial numbers of susceptible and infected individuals, and a fixed value of R_0, it is computationally possible to estimate the number of susceptible individuals at the end of the epidemic by solving the transcendental equation (2.11) for u_∞.

Example 2.5 (Linear cascade in predator–prey interactions) In many natural ecosystems, there are frequent interactions between predators and preys. Predators are members of species that eat members of another species. For example, a fox is a predator that eats rabbits. In this context, the rabbit is called prey. In the process of predation, predators often do not eat the entire prey. In ecology, this is considered as a wasteful killing. Here, we provide a classical example of a predator–prey model using a linear cascade of compartments (Figure 2.8).

Let x and y represent the sizes of prey and predator populations, respectively. We assume that the prey population grows according to the logistic model with the growth rate r and carrying capacity K. Suppose predators capture preys at the rate a, and eat them at the rate b. The difference between rates a and b may correspond to the phenomenon of wasteful killing. Assuming that predators die at the rate c, the dynamics of interactions can be expressed by the following system of differential equations [20]:

$$x' = rx\left(1 - \frac{x}{K}\right) - axy,$$
$$y' = bxy - cy.$$

Figure 2.8 Linear cascade of compartmental modelling for the dynamics of predator–prey interactions.

Similar to the model of disease epidemic discussed above, we cannot explicitly solve the system. However, the analysis tools from dynamical systems theory can help us explore the behavior of the solutions. We will discuss these tools in the next chapter and apply them to several mathematical models to investigate their dynamics.

Example 2.6 Branching cascade in epidemic dynamics) Communicable diseases often have different clinical manifestations. For example, influenza can be symptomatic with a number of different symptoms, including fever, cough, sore throat, runny or stuffy nose, muscle or body aches, and headaches. Influenza can also be asymptomatic, that is, without presenting any symptoms. A simplified dynamics of influenza spread among individuals can be shown by a branching cascade of compartments (Figure 2.9).

Suppose β_1 and β_2 represent the transmission rates of symptomatic and asymptomatic infected individuals, respectively. Let γ_1 and γ_2 be the corresponding recovery rates. We note that both symptomatically and asymptomatically infected individuals can transmit the disease to susceptible individuals. Suppose a fraction p of newly infected individuals will be symptomatic, and therefore the remaining fraction $(1 - p)$ will be asymptomatic. Thus, the dynamics of infection can be described by the following system of differential equations:

$$
\begin{aligned}
S' &= -\beta_1 SI - \beta_2 SA, \\
I' &= p(\beta_1 SI + \beta_2 SA) - \gamma_1 I, \\
A' &= (1 - p)(\beta_1 SI + \beta_2 SA) - \gamma_2 A, \\
R' &= \gamma_1 I + \gamma_2 A.
\end{aligned}
\tag{2.12}
$$

This branching cascade makes the model more complex than the linear cascade of epidemic discussed in Example 2.4.

2.2 Parameter Units

Checking the validity of model equations is one of the key steps in the modelling process. Often models include several parameters that describe the relationship

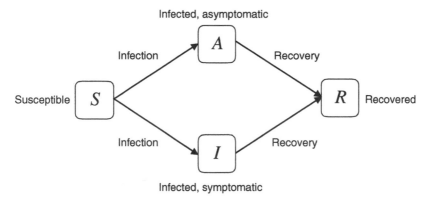

Figure 2.9 Branching cascade of compartmental modelling for the dynamics of influenza spread in a population.

between variables of the system and have specific units. Scaling can reduce the number of parameters and nondimensionalize the model. However, it is not always possible to find the appropriate change of variables for scaling. Furthermore, it is important to validate the model formulation before scaling or further analysis, in order to describe the system dynamics.

The values of parameters with dimensional quantity in a model are generally measured as some multiple of basic units. For example, in the logistic equation, $N(t)$ is the population size at any point in time, and therefore is in units of "people". The rate of change in the population size (i.e., the left-hand side of the equation), denoted by $\frac{dN}{dt}$, is in units of $\frac{people}{time}$. In order for the model to be mathematically valid, the right-hand side of the logistic equation (2.5) must have the same units. Expanding the right-hand side, we have two terms in the expression $rN - rN^2/K$. In the modelling process, adding terms is mathematically valid only if all terms have the same dimensions. Therefore, in the logistic equation, rN and rN^2/K are also in units of $\frac{people}{time}$. Multiplication of units in modelling is allowed regardless of the dimensions of terms. Since N is in units of "people", it follows that the growth rate is in units of $\frac{1}{time}$. For the second term, we note that the carrying capacity is the limiting size of the population, and therefore has the same dimensions as N. Thus, adding the two terms is valid and has the same dimensions as dN/dt.

In the process of validating the model for its dimensions, it is also possible to determine the parameter units. For example, we may be interested in determining the units of the transmission rate of a disease. Suppose the disease dynamics is governed by the model equations (2.7). The left-hand side of this equation is in units of $\frac{people}{time}$. The nonlinear terms on the right-hand side include the multiplication SI in units of "people^2". For this model to be mathematically valid, the transmission rate β must be in units of $\frac{1}{people \times time}$.

Exercises

1 For the logistic equation in (2.5), propose a modification so that the population goes extinct whenever N is below a certain value (referred to as the threshold of extinction).

2 The population of the USA in 1800 and 1850 was 5.3 and 23.1 million people, respectively. What is the predicted population of the USA in 1900 and 1950 using the exponential growth model? The actual population of the USA in 1900 was 76.2 million people. Suppose the growth rate is 0.0315 in the logistic equation. Determine the carrying capacity to predict the US population in 1950.

3 In the logistic equation, suppose $N_0 < K/2$. Determine the time at which the population size reaches $K/2$.

4 **Bacterial growth in a chemostat.** The chemostat is a continuous-flow culture, which is widely used in the study of microbial physiology, such as bacterial growth [37]. Within this controlled condition, the bacteria grow in the main culture vessel which contains a fresh nutrient culture medium that is continuously supplied at a constant flow rate of n. The overflow is also continuously removed at the same rate to maintain the culture volume at a constant v at all times. A compartmental modelling framework for this system is given by:

$$\frac{dW}{dt} = rW - \left(\frac{n}{v}\right)W,$$

$$\frac{dN}{dt} = nc_r - \left(\frac{n}{v}\right)N - \delta rW,$$

where W and N are the concentrations of bacteria and nutrients, r is the per capita reproduction number, n/v is the dilution rate, c_r is the supply nutrient concentration, and δ represents the stoichiometry of nutrient conversion to growth. Describe this model in a linear compartmental cascade. Determine the units of each parameter. Using the change of variable $C = N/v$, obtain the corresponding equation for the nutrient concentration. In the next chapter, we will learn about analysis of models for their critical points. The stability analysis techniques can be used to analyze the steady state of this chemostat model.

5 **The Monod growth model.** A similar model of growth to the chemostat was proposed by J. Monod based on a series of experiments in the

1930s (published in 1942) [34]. In this model, as the number of nutrients increases, the growth rate of the bacteria increases and eventually saturates. This saturation shows that after reaching some level of growth, an increase in the number of nutrients has little effect on the growth of the bacteria. A Monod-type model can be expressed by:

$$\frac{dB}{dt} = \left(\frac{rN}{N+M}\right)B - \delta B,$$
$$\frac{dN}{dt} = aN - \left(\frac{brN}{N+M}\right)B,$$

where B and N are the concentrations of the bacteria and nutrient, respectively; r and δ are the growth and death rates, respectively, of the bacteria; a is the rate of nutrient supply, b is the conversion rate of nutrient to the bacterial growth, and M is the saturation effect parameter. Describe this model in a compartmental cascade. Determine the units of each parameter. This model can also be analyzed using techniques proposed in the next chapter.

6 **Disease spread model.** Let us revisit the model in Example 2.6, and expand it to include a population of exposed individuals (E) who are infected but not yet infectious. The dynamics of the model, schematically illustrated in the figure below, is governed by the following differential equations:

$$S' = -\beta_I IS - \beta_A AS + \delta R,$$
$$E' = +\beta_I IS + \beta_A AS - \sigma E,$$
$$A' = \alpha \sigma E - d_A A,$$
$$I' = (1 - \alpha)\sigma E - d_I I,$$
$$R' = d_A A + d_I I - \delta R.$$

Suppose β_I and β_A represent the transmission rates of symptomatically and asymptomatically infected individuals, respectively. Let σ represent the rate at which exposed individuals become infectious; and d_I and d_A be the recovery rates of symptomatic and asymptomatic infections, respectively. We denote the rate at which recovered individuals lose their immunity and become susceptible again by δ. The fraction of exposed individuals who develop a symptomatic infection is denoted by α.

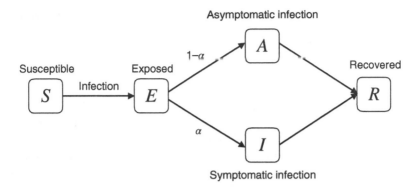

Asymptomatic infection

At any point of time, the rate of new infections (i.e., incidence) for symptomatic and asymptomatic cases can be calculated by $J_I = \beta_I I S$ and $J_A = \beta_A A S$, respectively. The proportion of the total incidence (P) attributable to asymptomatic cases is given by:

$$P = \frac{\int_0^\infty J_A(t)dt}{\int_0^\infty (J_A(t) + J_I(t))dt}.$$

Show that, for $0 < \alpha < 1$,

$$P = \frac{\alpha r}{1 + \alpha(r - 1)},$$

where

$$r = \frac{\beta_A d_A}{\beta_I d_I}.$$

7 Consider the following differential equation:

$$N' = rN \left(1 - \frac{N}{K}\right)\left(1 - \frac{K}{2N}\right),$$

where r and K are positive constants. Use an appropriate change of variables to scale this equation to an equation with dimensionless independent and dependent variables.

8 Consider the following two-compartment model which illustrates drug disposition from a network of capillaries to tissue compartment:

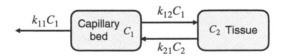

The system dynamics is governed by the following differential equations:

$$\frac{dC_1}{dt} = k_{21}C_2 - k_{12}C_1 - k_{11}C_1,$$

$$\frac{dC_2}{dt} = k_{12}C_1 - k_{21}C_2,$$

where C_1 and C_2 represent concentrations of drug in capillary bed and tissue compartments, respectively; k_{12} and k_{21} are the transition rates between these compartments; and k_{11} denotes the constant drug elimination rate. Solve the system and show that these concentrations (i.e., the solutions) follow a combination of exponential terms.

To find the transfer rate constants for this two-compartment model, the concentration of drug in the first compartment must be given by:

$$C_1 = \text{Distribution} + \text{Elimination}.$$

Show that:

$$C_1 = \alpha e^{-At} + \beta e^{-Bt},$$

where (α, A) and (β, B) are constants for the two exponential terms related to distribution and elimination, respectively. When the initial decline due to distribution is faster than the final decline due to elimination of drug, show that:

$$\ln C_1(t) = \ln \beta - \frac{Bt}{2.303}.$$

3

Analysis Tools

Mathematical models are often complex and nonlinear, and therefore cannot be solved to obtain explicit solutions. Dynamical systems theory provides important theoretical tools to study the qualitative behavior of such models without the need to explicitly formulate their solutions in terms of variables and parameters [39]. In this chapter, we consider a number of analytical tools and present examples of mathematical models to describe the utility of such tools in the analysis of dynamical systems.

In order to introduce these tools, we consider a general class of ordinary differential equations in a real n-dimensional space ($\mathbb{R}^n$). Suppose a system has n variables defined by $X = (x_1, x_2, \ldots, x_n)$, with the following relationship in time t:

$$X'(t) = F(x_1(t), x_2(t), \ldots, x_n(t)), \tag{3.1}$$

where F is a continuously differentiable function, with parameters (not shown) describing the relation among variables $x_1, x_2, \ldots, x_n$.

Definition 3.1 Any real solution $X^* = (x_1^*, x_2^*, \ldots, x_n^*)$ of (3.1) at which

$$F(x_1^*, x_2^*, \ldots, x_n^*) = 0, \quad \text{for all } t,$$

is called a *critical point* of the system.

In general, the system behavior with respect to a critical point can be analyzed locally or globally. In local analysis, the system dynamics is investigated in a neighborhood of the critical points in $\mathbb{R}^n$. In global analysis, the system dynamics is explored within the entire domain of its variables. In this book, we restrict our attention to the local analysis of mathematical models around their critical points.

Definition 3.2 A critical point X^* is called *isolated* if there exists a nonempty neighborhood around X^* in which the system has no other critical points.

Mathematical Modelling: A Graduate Textbook, First Edition. Seyed M. Moghadas and Majid Jaberi-Douraki.
© 2019 John Wiley & Sons, Inc. Published 2019 by John Wiley & Sons, Inc.
Companion Website: www.wiley.com/go/Moghadas/Mathematicalmodelling

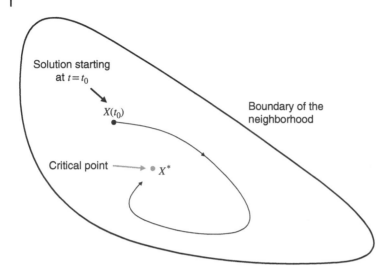

Figure 3.1 Representation of the stable node X^* in a neighborhood.

Example 3.1 Consider the following system of differential equations in the (x, y) plane:

$$x' = y,$$
$$y' = -x.$$

(3.2)

This system has a unique critical point $(0, 0)$, which is isolated.

Definition 3.3 An isolated critical point X^* is called a *stable node* if there exists a neighborhood of X^* such that (Figure 3.1):

1) Every solution of the system starting at time $t = t_0$ in this neighborhood is defined for all $t > t_0$.
2) Every solution of the system starting in this neighborhood remains in this neighborhood for all $t > t_0$.

Definition 3.4 A stable node X^* is *locally asymptotically stable* if there exists a neighborhood such that, for every solution $X(t)$ starting in this neighborhood, we have:

$$\lim_{t \to \infty} X(t) = X^*.$$

Question: Is the critical point $(0, 0)$ in system (3.2) a locally asymptotically stable node?

To answer this question, we solve the system to obtain the general form of its solutions. These solutions can be obtained by taking the second derivative

Figure 3.2 Representation of the solutions for system (3.2) as circles around the origin.

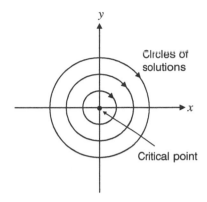

Circles of solutions

Critical point

of x and using the system to obtain the equation $x'' + x = 0$. The characteristic equation for this homogeneous equation has the solution $\lambda = \pm i$, giving the solutions:

$$x(t) = r\sin(t + \alpha),$$
$$y(t) = r\cos(t + \alpha),$$

where r and α are constants. Thus, $x^2 + y^2 = r^2$, indicating that the solutions are circles around the origin with the radius r (Figure 3.2). This means that for any neighborhood around $(0,0)$, there exists a positive $r > 0$ such that the solution $x^2 + y^2 = r$ remains in this neighborhood, but

$$\lim_{t \to \infty} x(t) \neq 0, \quad \lim_{t \to \infty} y(t) \neq 0.$$

Hence, $(0,0)$ is not locally asymptotically stable in system (3.2), while it is a stable node.

Example 3.2 Consider the following system of differential equations:

$$x' = -x,$$
$$y' = -y.$$
(3.3)

In this system, $(0,0)$ is the only critical point, and it is isolated. Solving the system gives the solutions:

$$x(t) = C_1 e^{-(t-t_0)},$$
$$y(t) = C_2 e^{-(t-t_0)},$$

where C_1 and C_2 are constants. We note that:

$$\lim_{t \to \infty} x(t) = \lim_{t \to \infty} y(t) = 0.$$

Assuming that $C_1 \neq 0$, we get:

$$y(t) = \frac{C_2}{C_1} x(t),$$

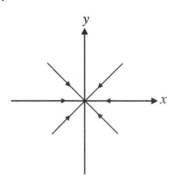

Figure 3.3 Representation of the solutions of (3.3) approaching the stable node (0, 0) as t goes to ∞.

which represents a line in the (x, y) plane passing through the origin (Figure 3.3). Therefore, $(0, 0)$ is a stable node which is also locally asymptotically stable.

3.1 Stability Analysis

In previous examples, we were able to find the general form of the solutions for the systems, which enabled us to understand the local behavior of the solutions around the critical points. However, this may not be an easy task in more complex models. Naturally, a question arises as to how one can determine the type of critical points and local dynamics of a system whose solutions cannot be explicitly obtained. To answer this question, let us consider the system:

$$X'(t) = F(X(t)), \tag{3.4}$$

where $X(t) = (x_1(t), x_2(t), \ldots, x_n(t))$, $F(X)$ is differentiable and its derivative with respect to each variable x_i $(i = 1, \ldots, n)$ is continuous. We rewrite the system in the form:

$$x'_1 = f_1(x_1, x_2, \ldots, x_n),$$
$$x'_2 = f_2(x_1, x_2, \ldots, x_n),$$
$$\vdots \tag{3.5}$$
$$x'_n = f_n(x_1, x_2, \ldots, x_n).$$

Suppose $X^* = (x_1^*, x_2^*, \ldots, x_n^*)$ is a critical point of (3.5). Thus,

$$f_1(x_1^*, \ldots, x_n^*) = f_2(x_1^*, \ldots, x_n^*) = \cdots = f_n(x_1^*, \ldots, x_n^*) = 0. \tag{3.6}$$

Let $X(t) = (x_1(t) + x_1^*, \ldots, x_n(t) + x_n^*)$ be a solution of (3.5) in a small neighborhood of X^*. This is called a perturbed solution. Expanding f_i using Taylor series around X^*, we get:

$$x_1' = f_1(x_1^*, x_2^*, \ldots, x_n^*) + \left.\frac{\partial f_1}{\partial x_1}\right|_{X^*} x_1 + \cdots + \left.\frac{\partial f_1}{\partial x_n}\right|_{X^*} x_n + O(x^2),$$

$$\vdots$$

$$x_n' = f_n(x_1^*, x_2^*, \ldots, x_n^*) + \left.\frac{\partial f_n}{\partial x_1}\right|_{X^*} x_1 + \cdots + \left.\frac{\partial f_n}{\partial x_n}\right|_{X^*} x_n + O(x^2),$$

where $O(x^2)$ represents terms of order 2 and higher. Since X^* is a critical point, (3.6) holds and we can write the system in a matrix form:

$$X' = \begin{bmatrix} \dfrac{\partial f_1}{\partial x_1} & \cdots & \dfrac{\partial f_1}{\partial x_n} \\ \vdots & \ddots & \vdots \\ \dfrac{\partial f_n}{\partial x_1} & \cdots & \dfrac{\partial f_n}{\partial x_n} \end{bmatrix}_{X^*} X + O(X^2),$$

or simply $X' = J_{X^*}X + O(X^2)$. The matrix J_{X^*} is called the *Jacobian* of (3.5) at X^*, and $X' = J_{X^*}X$ represents the corresponding linearized system. The eigenvalues of J_{X^*} are solutions (λ) of the characteristic equation:

$$\det(J_{X^*} - \lambda I) = 0.$$

This characteristic equation is a polynomial of degree n (with real coefficients), and therefore has n real or complex roots.

Definition 3.5   An isolated critical point X^* of (3.5) is called *hyperbolic* if all the eigenvalues of J_{X^*} have nonzero real parts.

Theorem 3.1 (Hartman–Grobman theorem)   Suppose X^* is a hyperbolic critical point of the system (3.5). Then (3.5) has the same qualitative structure (i.e., dynamical behavior) as the linearized system $X' = J_{X^*}X$ in a neighborhood of X^*.

A proof of the Hartman–Grobman theorem can be found in [39]. Here we apply this theorem to analyze the local dynamics of mathematical models around their critical points.

Definition 3.6   Suppose X^* is a hyperbolic isolated critical point of the system (3.5).

1) X^* is called a *sink* (stable node) if all the eigenvalues of J_{X^*} have negative real parts.
2) X^* is called a *source* (unstable node) if all the eigenvalues of J_{X^*} have positive real parts.
3) X^* is called a *saddle node* (unstable node) if J_{X^*} has at least one eigenvalue with positive real part and one eigenvalue with negative real part.

Definition 3.7 Suppose $X' = F(X)$ is a system with the above properties. A subset S of $\mathbb{R}^n$ ($S \subseteq \mathbb{R}^n$) is called *positively invariant* if, for any solution $X(t)$ starting in S at time $t = 0$, $X(t) \subseteq S$ for all $t > 0$ (i.e., X remains in S for $t > 0$). The subset S is called *negatively invariant* if, for any solution $X(t)$ starting in S at time $t = 0$, $X(t) \subseteq S$ for all $t < 0$ (i.e., X remains in S for $t < 0$).

Definition 3.8 (Stable Manifold) For a hyperbolic critical point X^* of the system $X' = F(X)$, the local stable manifold is a subset $E^s \subseteq \mathbb{R}^n$ such that E^s is positively invariant in a neighborhood of X^*. The union $\bigcup_{t \geq 0} E^s$ of all local stable manifolds is a global stable manifold.

Definition 3.9 (Unstable Manifold) For a hyperbolic critical point X^* of the system $X' = F(X)$, the local unstable manifold is a subset $E^u \subseteq \mathbb{R}^n$, such that E^u is negatively invariant in a neighborhood of X^*. The union $\bigcup_{t \leq 0} E^u$ of all local unstable manifolds is a global unstable manifold.

Theorem 3.2 (Local center manifold theorem) Let X^* be a hyperbolic critical point of the system $X' = F(X)$ ($X \in \mathbb{R}^n$) with the above assumptions. Suppose J_{X^*} has m eigenvalues with negative real parts, k eigenvalues with positive real parts, and $j = n - m - k$ eigenvalues with zero real parts. Then the system has a j-dimensional center manifold E^c, an m-dimensional stable manifold E^s, and a k-dimensional unstable manifold E^u.

A proof of the center manifold theorem can be found in [39]. We denote the dimension of stable, unstable, and center manifolds by $\dim(E^s)$, $\dim(E^u)$, and $\dim(E^c)$, respectively. Note that $\dim(E^s) + \dim(E^u) + \dim(E^c) = n$.

Example 3.3 To apply the above concepts, let us consider the following predator–prey system:

$$x' = x(1 - x) - 2xy,$$
$$y' = (x - 5)y.$$

The critical points are $(0, 0), (1, 0), (5, -2)$. The Jacobian matrix associated with the linearized system is:

$$J = \begin{bmatrix} 1 - 2x - 2y & -2x \\ y & x - 5 \end{bmatrix}.$$

At $X_1 = (0, 0)$, the Jacobian is:

$$J_1 = \begin{bmatrix} 1 & 0 \\ 0 & -5 \end{bmatrix},$$

and the eigenvalues are given by diagonal entries $\lambda_1 = 1$ and $\lambda_2 = -5$. This shows that the critical point X_1 is a saddle node. Furthermore, $\dim(E^s) = \dim(E^u) = 1$ and $\dim(E^c) = 0$.

At $X_2 = (1, 0)$, the Jacobian is:

$$J_2 = \begin{bmatrix} -1 & -2 \\ 0 & -4 \end{bmatrix},$$

and the eigenvalues are given by diagonal entries $\lambda_1 = -1$ and $\lambda_2 = -4$. This shows that the critical point X_2 is a sink (stable node). Furthermore, $\dim(E^s) = 2$ and $\dim(E^u) = \dim(E^c) = 0$.

Finally, at $X_3 = (5, -2)$, the Jacobian is:

$$J_3 = \begin{bmatrix} -5 & -10 \\ -2 & 0 \end{bmatrix},$$

with the characteristic equation $\lambda^2 + 5\lambda - 20 = 0$. The eigenvalues are:

$$\lambda = \frac{-5 + \sqrt{105}}{2} > 0,$$

$$\lambda = \frac{-5 - \sqrt{105}}{2} < 0,$$

and therefore X_3 is a saddle node. Thus, $\dim(E^s) = \dim(E^u) - 1$ and $\dim(E^c) = 0$.

Example 3.4 Consider the following system of differential equations:

$$\frac{dx}{dt} = a + rx\left(1 - \frac{x}{K}\right) - bxy,$$

$$\frac{dy}{dt} = cyz - dy,$$

$$\frac{dz}{dt} = sz\left(1 - \frac{z}{L}\right) - cyz - ez, \tag{3.7}$$

where all the parameters are positive. We now find all critical points with at least one positive component, and determine their stability. Setting the right-hand side of the system equal to zero and solving the equations gives the critical points.

When $y = z = 0$, we find the critical point:

$$C_1 = \left(\frac{K}{2}\left(1 + \sqrt{1 + \frac{4a}{rK}}\right), 0, 0 \right).$$

The Jacobian of the system at C_1 is:

$$J_{C_1} = \begin{bmatrix} -r\sqrt{1 + \frac{4a}{rK}} & \frac{-bK}{2}\left(1 + \sqrt{1 + \frac{4a}{rK}}\right) & 0 \\ 0 & -d & 0 \\ 0 & 0 & s - e \end{bmatrix}.$$

Since:

$$\lambda_1 = -r\sqrt{1 + \frac{4a}{rK}} < 0, \quad \lambda_2 = -d < 0,$$

we conclude that C_1 is a stable node if $\lambda_3 = s - e < 0$, and an unstable node if $s - e > 0$. Thus, we have:

$$\dim(E^s) = 3, \ \dim(E^u) = 0, \ \dim(E^c) = 0, \ \text{if } s - e < 0,$$
$$\dim(E^s) = 2, \ \dim(E^u) = 1, \ \dim(E^c) = 0, \ \text{if } s - e > 0,$$
$$\dim(E^s) = 2, \ \dim(E^u) = 0, \ \dim(E^c) = 1, \ \text{if } s - e = 0.$$

When $s - e > 0$, the second positive critical point of the system is:

$$C_2 = \left(\frac{K}{2} \left(1 + \sqrt{1 + \frac{4a}{rK}} \right), 0, L \left(\frac{s - e}{s} \right) \right).$$

The Jacobian at C_2 is:

$$J_{C_2} = \begin{bmatrix} -r\sqrt{1 + \frac{4a}{rK}} & \frac{-bK}{2} \left(1 + \sqrt{1 + \frac{4a}{rK}} \right) & 0 \\ 0 & \frac{cL(s-e)}{s} - d & 0 \\ 0 & -\frac{cL(s-e)}{s} & -(s - e) \end{bmatrix}.$$

Since:

$$\lambda_1 = -r\sqrt{1 + \frac{4a}{rK}} < 0, \quad \lambda_3 = -(s - e) < 0,$$

the stability of C_2 depends on the sign of:

$$\lambda_2 = \frac{cK(s - e)}{s} - d.$$

Thus, we have:

$$\dim(E^s) = 3, \ \dim(E^u) = 0, \ \dim(E^c) = 0, \ \text{if } \frac{cL(s-e)}{sd} < 1,$$
$$\dim(E^s) = 2, \ \dim(E^u) = 1, \ \dim(E^c) = 0, \ \text{if } \frac{cL(s-e)}{sd} > 1,$$
$$\dim(E^s) = 2, \ \dim(E^u) = 0, \ \dim(E^c) = 1, \ \text{if } \frac{cL(s-e)}{sd} = 1.$$

When $\frac{cL(s-e)}{sd} > 1$, the system exhibits another critical point, given by:

$$C_3 = \left(A, \frac{s}{c} \left(1 - \frac{d}{cL} \right) - \frac{e}{c}, \frac{d}{c} \right),$$

where A is the solution of the quadratic:

$$\frac{r}{K} A^2 + \left(\frac{bs}{c} \left(1 - \frac{d}{cL} \right) - \frac{be}{c} - r \right) A - a = 0.$$

The Jacobian at C_3 is:

$$J_{C_3} = \begin{bmatrix} -\frac{a}{A} - \frac{rA}{K} & -bA & 0 \\ 0 & 0 & s\left(1 - \frac{d}{cL}\right) - e \\ 0 & -d & -\frac{sd}{cL} \end{bmatrix}.$$

From the condition $\frac{cL(s-e)}{sd} > 1$, it follows that all the eigenvalues of J_{C_3} have negative real parts and therefore C_3 is a stable node with:

$$\dim(E^s) = 3, \ \dim(E^u) = 0, \ \dim(E^c) = 0.$$

3.2 Phase-Plane Behavior

In this section we analyze the local dynamics in a two-dimensional space ($\mathbb{R}^2$). Let us consider the system:

$$x' = f_1(x, y),$$
$$y' = f_2(x, y), \tag{3.8}$$

where f_1 and f_2 are differentiable functions with respect to x and y, and their derivatives are continuous. Let $X^* = (x^*, y^*)$ be a an isolated critical point of the system. The linearized system around X^* has the form:

$$\begin{bmatrix} x' \\ y' \end{bmatrix} = \begin{bmatrix} \frac{\partial f_1}{\partial x} & \frac{\partial f_1}{\partial y} \\ \frac{\partial f_2}{\partial x} & \frac{\partial f_2}{\partial y} \end{bmatrix}_{X^*} \begin{bmatrix} x \\ y \end{bmatrix}.$$

If we write the Jacobian matrix at X^* in the form:

$$J_{X^*} = \begin{bmatrix} a_{11} & a_{12} \\ a_{21} & a_{22} \end{bmatrix},$$

then the characteristic equation of J_{X^*} is:

$$\lambda^2 - (a_{11} + a_{22})\lambda + a_{11}a_{22} - a_{12}a_{21} = 0.$$

Let $p = a_{11} + a_{22}$ and $q = a_{11}a_{22} - a_{12}a_{21}$. Thus, the eigenvalues are:

$$\lambda_1 = \frac{p + \sqrt{\Delta}}{2}, \quad \lambda_2 = \frac{p - \sqrt{\Delta}}{2},$$

where $\Delta = p^2 - 4q$. We now consider the following three cases.

1) $q < 0$. In this case, $\sqrt{\Delta} > |p| > 0$, and therefore $\lambda_1 > 0$ and $\lambda_2 < 0$. Thus, X^* is a saddle node with $\dim(E^s) = \dim(E^u) = 1$, and $\dim(E^c) = 0$. (Figure 3.4) Note that the dimension of the system is $\dim(E^s) + \dim(E^u) + \dim(E^c)$.

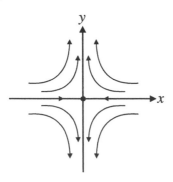

Figure 3.4 Representation of the local dynamics of a saddle node in the (x, y) plane.

(a)   (b)

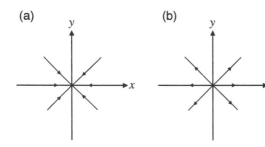

Figure 3.5 Representation of the local dynamics of (a) a stable node (sink); and (b) an unstable node, in the (x, y) plane.

2) $0 \leq q \leq p^2/4$. In this case, $\Delta > 0$ and $\sqrt{\Delta} < |p|$. Hence, we have two possibilities:
   a) If $p < 0$, then $\lambda_1 < 0$ and $\lambda_2 < 0$. Thus, X^* is locally asymptotically stable, with $\dim(E^s) = 2$ and $\dim(E^u) = \dim(E^c) = 0$ (Figure 3.5a).
   b) If $p > 0$, then $\lambda_1 > 0$ and $\lambda_2 > 0$. Thus, X^* is an unstable node, with $\dim(E^u) = 2$ and $\dim(E^s) = \dim(E^c) = 0$ (Figure 3.5b).
3) $q > p^2/4$ and $\Delta < 0$. In this case, the two eigenvalues are complex numbers, and we have three possibilities:
   a) If $p < 0$, then both eigenvalues have negative real parts and therefore X^* is a stable (spiral) node. (Figure 3.6a).
   b) If $p > 0$, then both eigenvalues have positive real parts and therefore X^* is an unstable (spiral) node. (Figure 3.6b).
   c) If $p = 0$, then both eigenvalues are purely imaginary $\lambda_{1,2} = \pm i\sqrt{\Delta}$. Thus, X^* is a center and referred to as *neutral* critical point (Figure 3.7).

The local dynamics of the system (3.8) in a two-dimensional space is summarized in Figure 3.8.

Figure 3.6 Representation of the local dynamics of (a) a stable spiral node; and (b) an unstable spiral node, in the (x, y) plane

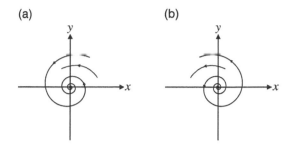

(a)   (b)

Figure 3.7 Representation of the local dynamics of a center in the (x, y) plane.

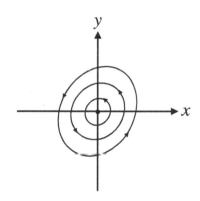

Example 3.5   Consider the following system of differential equations:

$$x' = xy + 3y,$$
$$y' = xy - 3x. \tag{3.9}$$

This system has two critical points, $(x, y) = (0, 0)$ and $(x, y) = (-3, 3)$. The Jacobian of (3.9) is:

$$J = \begin{bmatrix} y & x+3 \\ y-3 & x \end{bmatrix}.$$

At $(0, 0)$, the Jacobian is:

$$J = \begin{bmatrix} 0 & 3 \\ -3 & 0 \end{bmatrix},$$

with the characteristic equation $\lambda^2 + 9 = 0$. Thus, the eigenvalues are purely imaginary and given by $\lambda_{\pm} = \pm 3i$. Hence, $(0, 0)$ is a center and $\dim(E^c) = 2$.
At $(-3, 3)$, the Jacobian is:

$$J = \begin{bmatrix} 3 & 0 \\ 0 & -3 \end{bmatrix},$$

with the eigenvalues $\lambda_{\pm} = \pm 3$. Thus, $(-3, 3)$ is a saddle node, with $\dim(E^s) = \dim(E^u) = 1$. Figure 3.9 shows the phase plane of the system (3.9).

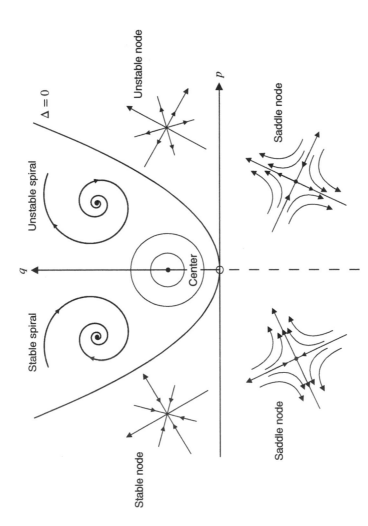

Figure 3.8 Representation of the dynamical behavior of system (3.8) around its critical points based on the constant values of p and q.

Figure 3.9 Representation of the phase plane for system (3.9).

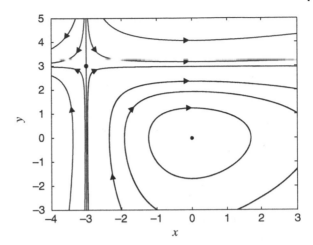

Example 3.6 (Gause-type predator–prey system) We consider a general form of the predator–prey system [15, 31] with the following system of differential equations in which the prey population grows according to the logistic model:

$$x' = rx\left(1 - \frac{x}{K}\right) - axy,$$
$$y' = (cx - d)y, \tag{3.10}$$

where r, K, a, c, and d are the parameters, assumed to be positive. It is easy to show that the critical points are:

$$X_1 = (0,0), \quad X_2 = (K,0), \quad X_3 = \left(\frac{d}{c}, \frac{r}{a}\left(1 - \frac{d}{cK}\right)\right).$$

The Jacobian of (3.10) has the form:

$$J = \begin{bmatrix} r\left(1 - \frac{2x}{K}\right) - ay & -ax \\ cy & cx - d \end{bmatrix}.$$

We consider a number of possibilities for each critical point to find the stability conditions.

1) At X_1, the Jacobian is:

$$J_{X_1} = \begin{bmatrix} r & 0 \\ 0 & -d \end{bmatrix},$$

with eigenvalues $\lambda_1 = r$ and $\lambda_2 = -d$. Thus, X_1 is a saddle node, with $\dim(E^s) = \dim(E^u) = 1$ and $\dim(E^c) = 0$. Considering a solution

$X(t) = (x(t), y(t))$ in a small neighborhood of X_1, we have the following cases:

a) If $0 < x < K$ and $y = 0$, then $x' > 0$ and x increases. If $x < 0$ and $y = 0$, then $x' < 0$ and x decreases. Thus, the x-axis provides the unstable manifold E^u of the system at X_1 and $\lim_{t \to -\infty} x(t) = 0$.

b) If $y > 0$ and $x = 0$, then $y' < 0$ and therefore y decreases. If $y < 0$ and $x = 0$, then $y' > 0$ and y increases. This means that the y-axis provides the stable manifold E^s at X_1 and $\lim_{t \to \infty} y(t) = 0$.

2) At X_2, the Jacobian is:

$$J_{X_2} = \begin{bmatrix} -r & -aK \\ 0 & cK - d \end{bmatrix},$$

with eigenvalues $\lambda_1 = -r$ and $\lambda_2 = cK - d$. Therefore, we have the following possibilities:

a) If $cK - d > 0$, then X_2 is a saddle node with $\dim(E^s) = \dim(E^u) = 1$ and $\dim(E^c) = 0$. Similar to the argument for X_1, one can show that the x-axis is the stable manifold E^s of the system at X_2. The unstable manifold E^u is generated by the solution curves that approach $(K, 0)$ as $t \to -\infty$ when $x \neq 0$.

b) If $cK - d < 0$, then X_2 is a stable node with $dim(E^s) = 2$, and $\dim(E^u) = \dim(E^c) = 0$. In this case, the x-axis provides one dimension of E^s and solution curves that approach $(K, 0)$ as $t \to \infty$ with $x \neq 0$ provide another dimension of E^s.

Note that if $cK - d = 0$, then $X_2 = X_3$. In this case, X_2 has a one-dimensional center manifold. Computationally, it can be shown that X_2 is stable for $y > 0$ and attracts the solutions starting with initial values that have $y > 0$ (Figure 3.10). In the remaining part of our analysis, we assume that $cK - d \neq 0$.

3) At X_3, the Jacobian is:

$$J_{X_3} = \begin{bmatrix} -\dfrac{rd}{cK} & -\dfrac{ad}{c} \\ \dfrac{rc}{a}\left(1 - \dfrac{d}{cK}\right) & 0 \end{bmatrix},$$

with the characteristic equation:

$$\lambda^2 + \frac{rd}{cK}\lambda + rd\left(1 - \frac{d}{cK}\right) = 0.$$

The eigenvalues are:

$$\lambda_1 = \frac{-\dfrac{rd}{cK} + \sqrt{\Delta}}{2}, \quad \lambda_2 = \frac{-\dfrac{rd}{cK} - \sqrt{\Delta}}{2},$$

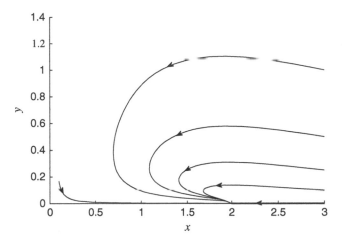

Figure 3.10 Solutions of the system (3.10) approaching X_2 when $cK - d = 0$. Parameter values are $r = 0.6, a = 1, K = 2, c = 1/2$, and $d = 1$.

where

$$\Delta = \left(\frac{rd}{cK}\right)^2 - 4rd\left(1 - \frac{d}{cK}\right).$$

There are two cases to consider.

a) If $cK - d > 0$, then X_3 is located in the $(x > 0, y > 0)$ region. Then $\lambda_1, \lambda_2 < 0$, and therefore X_3 is a locally asymptotically stable node with $\dim(E^s) = 2$ and $\dim(E^u) = \dim(E^c) = 0$.

b) If $cK - d < 0$, then X_3 is located in the $(x > 0, y < 0)$ region and $\Delta > 0$. While this critical point is biologically infeasible, its stability can be theoretically analyzed in a similar way to case (a), and we leave it as an exercise. A case in which the solutions starting in the $(x > 0, y < 0)$ region approach X_2 is illustrated in Figure 3.11.

Example 3.7 (A model for type 1 diabetes) Type 1 diabetes (T1D) is an autoimmune disease associated with the destruction of pancreatic β cells which generates a perturbation in insulin secretion. Different cytotoxic T-cells (i.e., immune responses), activated by peptides expressed on membranes of antigen presenting cells, are implicated in disease attack and progression. The dynamics of T1D can be described by the following system of differential equations:

$$\frac{dT}{dt} = (\sigma + \alpha T)\frac{P}{P + K} - \left(\sqrt{\alpha} - \sqrt{\delta_T}\right)^2 T^2 - \delta_T T,$$

$$\frac{dP}{dt} = \delta_P(T - P),$$

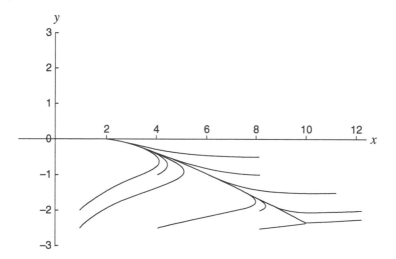

Figure 3.11 Solutions of the system (3.10) starting in the ($x > 0$, $y < 0$) region that approach X_2 when $cK - d < 0$. Parameter values are $r = 0.6$, $a = 1$, $K = 2$, $c = 0.1$, and $d = 1$.

where T represents the population of cytotoxic T-cells and P denotes the population of peptides expressed by antigen presenting cells which help activate T-cells. Also σ is the influx rate of naïve T-cells from the thymus, K represents the half-maximal level for activation of T-cells, α denotes the expansion rate of T-cells, and δ_T and δ_P denote the turnover rates of T-cells and peptides, respectively [22].

For this system, the origin $E^* = (0, 0)$ is a critical point. Also, depending on the parameter values, we may have at most two other positive critical points that can be found by solving the following equations:

$$P^* = T^*,$$

$$\frac{\sigma + \alpha T^*}{T^* + K} = \delta_T + \left(\sqrt{\alpha} - \sqrt{\delta_T} \right)^2 T^*.$$

The critical points can be calculated as a function of parameter K by:

$$T^*(K) = \frac{(\alpha - \delta_T - l^2 K) \pm \sqrt{(\alpha - \delta_T - l^2 K)^2 + 4(\sigma - \delta_T K)l^2}}{2l^2},$$

where $l = \sqrt{\alpha} - \sqrt{\delta_T} > 0$. If $(\alpha - \delta_T - l^2 K)^2 + 4(\sigma - \delta_T K)l^2 = 0$, then there is a unique critical point, and we have:

$$K_\pm = \frac{(\alpha + \delta_T) \pm 2\sqrt{\alpha \delta_T - \sigma l^2}}{l^2}.$$

If we ignore the thymus input by setting $\sigma = 0$, we obtain:

$$K_- - 1, \quad K_+ = \left(\frac{\sqrt{\alpha} + \sqrt{\delta_T}}{\sqrt{\alpha} - \sqrt{\delta_T}} \right)^2 .$$

The second case is impossible since $T^*(K_+) < 0$. It can also be seen that $T^*(K_-) = \sqrt{\delta_T}/l$. The Jacobian matrix has the following form:

$$J(T, P) = \begin{bmatrix} \frac{\alpha P}{P+K} - \delta_T - 2l^2 \, T & \frac{\alpha \, T}{P+K} - \frac{\alpha \, T \, P}{(P+K)^2} \\ \delta_P & -\delta_P \end{bmatrix} .$$

The eigenvalues of $J(T, P)$ at $E^* = (0, 0)$ are $\lambda_1 = -\delta_T$ and $\lambda_2 = -\delta_P$, and therefore E^* is a stable node. For positive critical points, we have:

$$\det(J(T, P)) = \delta_P l^2 T - \delta_P \left(\frac{\alpha T}{P + K} - \frac{\alpha T P}{(P + K)^2} \right)$$

$$= \delta_P \left(-\delta_T + \frac{\alpha T^2}{(T + K)^2} \right)$$

$$= \delta_P \left(-\delta_T + \frac{(\delta_T + l^2 T)^2}{\alpha} \right) .$$

Thus, $\det(J(T, P)) = 0$ when $T = T^*(K_-)$. This means that if $T > \sqrt{\delta_T}/l$, then $\det(J(T, P)) > 0$ and the critical point is a stable node, while the critical point is a saddle node for $T < \sqrt{\delta_T}/l$.

☞ **Liénard equation.** Suppose that f and g are continuously differentiable functions, $g(x)$ is an odd function (i.e., $g(-x) = -g(x)$), and $f(x)$ is an even function (i.e., $f(-x) = f(x)$). Then the system:

$$\frac{d^2 x}{dt^2} + f(x) \frac{dx}{dt} + g(x) = 0,$$

is called the Liénard equation [39]. This equation can be transformed to a system of two first-order ordinary differential equations by using the following change of variables:

$$F(x) = \int_0^x f(\xi) d\xi,$$

$$x_1 = x,$$

$$x_2 = x_1' + F(x_1).$$

Thus, using the chain rule, we get:

$$x_1' = x_2 - F(x_1),$$

$$x_2' = -g(x_1).$$

A particular example of the Liénard equation is the Van der Pol oscillator. In the relatively early days of the development of electronic technology, Van der Pol, a Dutch electrical engineer, investigated electrical circuits by employing vacuum tubes. He found that these circuits have stable oscillations behaving like periodic solutions. When the circuits are driven with a signal whose frequency is near the periodic solution, the resulting periodic response shifts its frequency to that of the driving signal. To describe these dynamics, Van der Pol proposed the following Liénard-type equation:

$$\frac{d^2x}{dt^2} - \mu(1-x^2)\frac{dx}{dt} + x = 0,$$

where $\mu \in \mathbb{R}$ is a constant parameter. Using the change of variables proposed in the Liénard equation, we get:

$$F(x) = \int_0^x -\mu(1-\xi^2)d\xi = -\mu\left(\xi - \frac{\xi^3}{3}\right)\Bigg|_0^x = -\mu\left(x - \frac{x^3}{3}\right).$$

Thus, the Van der Pol equation transforms into the following system of differential equations:

$$x_1' = x_2 + \mu\left(x_1 - \frac{x_1^3}{3}\right),$$

$$x_2' = -x_1.$$

This system has a unique critical point $(0,0)$. The Jacobian at $(0,0)$ is:

$$J_{(0,0)} = \begin{bmatrix} \mu & 1 \\ -1 & 0 \end{bmatrix},$$

with the characteristic equation $\lambda^2 - \mu\lambda + 1 = 0$. The eigenvalues are:

$$\lambda_1 = \frac{\mu + \sqrt{\mu^2 - 4}}{2}, \qquad \lambda_2 = \frac{\mu - \sqrt{\mu^2 - 4}}{2}.$$

We now consider the following possibilities.

1) If $\mu^2 - 4 \geq 0$, then $\mu \leq -2$ or $\mu \geq 2$, and $\sqrt{\mu^2 - 4} < |\mu|$. If $\mu \geq 2$, then $\lambda_1 > 0$ and $\lambda_2 > 0$, and hence $(0,0)$ is an unstable node. If $\mu \leq -2$, then $\lambda_1 < 0$ and $\lambda_2 < 0$, and therefore $(0,0)$ is a stable node.
2) If $\mu^2 - 4 < 0$, then $-2 < \mu < 2$. If $0 < \mu < 2$, then both eigenvalues are complex and have positive real parts. Thus $(0,0)$ is au unstable spiral node. If $-2 < \mu < 0$, then both eigenvalues are complex with negative real parts, and hence $(0,0)$ is a stable spiral node. If $\mu = 0$, then $\lambda_{1,2} = \pm 2i$ and therefore $(0,0)$ is a center.

In all examples considered here, the system has a finite number of critical points. We should point out that the system may have an infinite number of critical points. While this is beyond the scope of our analysis, we show this situation with two examples. Let us first consider the simple SI epidemic model:

$$S' - -\beta SI,$$
$$I' = \beta SI - \gamma I,$$

where $S(t)$ and $I(t)$ respectively represent the population of susceptible and infected individuals at time t. If $I = 0$, then $(S, 0)$ is a critical point of the system for any $S \in \mathbb{R}$. Clearly, these critical points are not isolated as any neighborhood of one critical point contains other critical points.

Let us now consider the following system:

$$x' = xy - x^2,$$
$$y' = xy - y^2. \tag{3.11}$$

It is easy to see that the line $y = x$ provides the set of all critical points, and therefore the system has no isolated critical point. Also, $\det(J) = 0$ along $y = x$.

3.3 Direction Field

Let us consider the general system:

$$x' = f_1(x, y),$$
$$y' = f_2(x, y),$$

where f_1 and f_2 are differentiable and their partial derivatives with respect to each variable are continuous. A solution passing through a point (x_0, y_0) in the domain of the variables at time t_0 has the following components of its tangent vectors:

$$f_1(x_0, y_0), \quad f_2(x_0, y_0).$$

The slope of the tangent line is given by $\frac{\Delta y}{\Delta x}$ (for sufficiently small line segment) or $\frac{dy}{dx}$ which represents $f_2(x, y)/f_1(x, y)$ at any given point in the (x, y) plane. A collection of such vectors defines a *direction field* of the system, which provides a visual guide in sketching a family of solution curves. A collective family of all solution curves is called the *phase plane*. The length of the tangent vector is $L = \sqrt{f_1^2 + f_2^2}$, but we do not attempt to accurately project the vectors in the (x, y) plane.

Example 3.8 Consider the following system of differential equations:

$$x' = 2 - x^2 - y^2,$$
$$y' = x^2 - y^2. \tag{3.12}$$

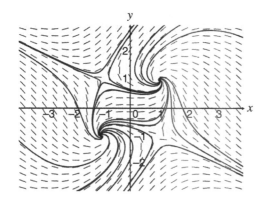

y

Figure 3.12 Direction field and phase plane for the system (3.12).

It is easy to see that the system has four critical points: $(1, 1)$, $(-1, 1)$, $(1, -1)$, $(-1, -1)$. The Jacobian of the system is:

$$J = \begin{bmatrix} -2x & -2y \\ 2x & -2y \end{bmatrix}.$$

At $(1, 1)$, the Jacobian has the characteristic equation $\lambda^2 + 4\lambda + 8 = 0$, with solutions $\lambda_\pm = -2 \pm 2i$. Thus, $(1, 1)$ is a stable node. At $(-1, -1)$, the characteristic equation is $\lambda^2 - 4\lambda + 8 = 0$, with solutions $\lambda_\pm = 2 \pm 2i$. Thus, $(-1, -1)$ is an unstable node. At $(-1, 1)$, the Jacobian has the characteristic equation $\lambda^2 - 8 = 0$, with solutions $\lambda_\pm = \pm 2\sqrt{2}$. Hence, $(-1, 1)$ is a saddle node. Similarly, it can be shown that $(1, -1)$ is a saddle node. The direction field and phase plane of system (3.12) are shown in Figure 3.12.

☞ **Isocline.** Consider the first-order differential equation:

$$\frac{dy}{dx} = f(x, y).$$

An isocline is a set of points in the direction field at which:

$$\frac{dy}{dx} = c,$$

for a constant $c \in \mathbb{R}$. Geometrically, the direction field vectors at the points of an isocline have the same slope. To find an isocline for a constant c, we need to solve $f(x, y) = c$.

Example 3.9 Consider the equation $y' = -x + y^3 - 2x^2$. Solving for the isoclines with $c = 20,\ 1, -10$, we obtain the curves shown in Figure 3.13. The vectors represent the direction field in the (x, y) plane.

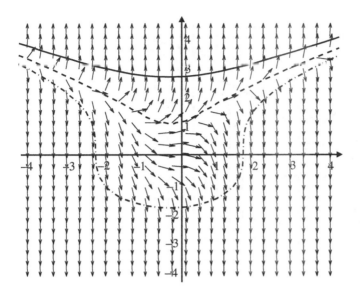

Figure 3.13 Direction field and isoclines for the equation $y' = -x + y^3 - 2x^2$ with $c = 20$ (solid curve) $c = 1$ (dashed curve), and $c = -10$ (dot-dashed curve).

☞ **Nullcline.** An isocline with constant $c = 0$ is called a nullcline. Geometrically, the direction field vectors are horizontal at these points.

Example 3.10 For the equation $y' = -x + y^3 - 2x^2$, we can solve $-x + y^3 - 2x^2 = 0$ to find the nullcline, shown by the bold curve in Figure 3.14. The nullcline corresponds to the curve $y = (x + 2x^2)^{\frac{1}{3}}$.

3.4 Round–Hurwitz Criterion

For systems defined on $\mathbb{R}^n$ with $n \leq 2$, stability analysis of the critical points depends on the eigenvalues of the characteristic equations which are the solutions of polynomials of degree $n \leq 2$. For systems in higher-dimensional spaces (i.e., $\mathbb{R}^n$ with $n \geq 3$), the characteristic equation is of degree $n \geq 3$ which may be difficult to solve. In this case, we may be able to use the Round–Hurwitz criterion [44] to determine whether the roots of such a polynomial have negative real parts [5].

Suppose X^* is a critical point of system (3.5) and J_{X^*} is the Jacobian at X^*. The characteristic equation is a polynomial of degree n, given by:

$$P(\lambda) = a_n \lambda^n + a_{n-1} \lambda^{n-1} + \cdots + a_1 \lambda + a_0.$$

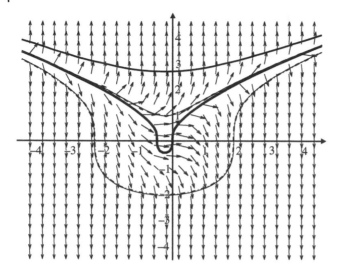

Figure 3.14 Nullcline (bold curve) for the equation $y' = -x + y^3 - 2x^2$ in Example 3.10. Other curves show the same isoclines illustrated in Figure 3.13.

The eigenvalues of J_{X^*} are the solutions of $P(\lambda) = 0$. Suppose the following conditions hold:

1) All coefficients of $P(\lambda)$ are nonzero.
2) All coefficients of $P(\lambda)$ have the same sign.

We now create the Routh–Hurwitz matrix in the form:

λ^n	a_n	a_{n-2}	a_{n-4}	a_{n-6}	$\cdots$
λ^{n-1}	a_{n-1}	a_{n-3}	a_{n-5}	a_{n-7}	$\cdots$
λ^{n-2}	b_1	b_2	b_3	b_4	$\cdots$
λ^{n-3}	c_1	c_2	c_3	c_4	$\cdots$
$\vdots$	$\vdots$	$\vdots$	$\vdots$	$\vdots$	$\cdots$
λ^0	q_1	q_2	q_3	q_4	$\cdots$

where the a_i are the coefficients of $P(\lambda)$ and other entries are calculated as follows:

$$b_1 = \frac{-1}{a_{n-1}} \det \begin{bmatrix} a_n & a_{n-2} \\ a_{n-1} & a_{n-3} \end{bmatrix}; \qquad c_1 = \frac{-1}{b_1} \det \begin{bmatrix} a_{n-1} & a_{n-3} \\ b_1 & b_2 \end{bmatrix}; \qquad \cdots$$

$$b_2 = \frac{-1}{a_{n-1}} \det \begin{bmatrix} a_n & a_{n-4} \\ a_{n-1} & a_{n-5} \end{bmatrix}; \qquad c_2 = \frac{-1}{b_1} \det \begin{bmatrix} a_{n-1} & a_{n-5} \\ b_1 & b_3 \end{bmatrix}; \qquad \cdots$$

$$b_3 = \frac{-1}{a_{n-1}} \det \begin{bmatrix} a_n & a_{n-6} \\ a_{n-1} & a_{n-7} \end{bmatrix}; \qquad c_3 = \frac{-1}{b_1} \det \begin{bmatrix} a_{n-1} & a_{n-7} \\ b_1 & b_4 \end{bmatrix}; \qquad \cdots$$

$$\vdots \qquad \vdots \qquad \vdots \qquad \vdots \qquad \vdots$$

According to the Routh–Hurwitz criterion [56], the necessary condition for all roots of $P(\lambda)$ (i.e., eigenvalues) to have negative real parts is that all elements of the first column in the Routh–Hurwitz matrix (i.e., $a_n, a_{n-1}, b_1, c_1, \ldots, q_1$) have the same sign. The number of changes that occur in the signs of these elements is equal to the number of roots with positive real parts.

☞ **Special case: $n = 3$.** Suppose the characteristic equation has the form:

$$P(\lambda) = \lambda^3 + a_2 \lambda^2 + a_1 \lambda + a_0 = 0.$$

The Routh–Hurwitz matrix is:

λ^3	1	a_1
λ^2	a_2	a_0
λ^1	$\dfrac{a_1 a_2 - a_0}{a_2}$	0
λ^0	a_0	

It follows from the conditions of the Routh–Hurwitz criterion that all the eigenvalues have negative real parts if:

$$a_0 > 0, \quad a_2 > 0, \quad a_1 a_2 > a_0.$$

Example 3.11 Consider $P(x) = x^3 + x^2 + 2x + 8$. To apply the Routh–Hurwitz criterion, we generate the following matrix:

$$
\begin{array}{c|cc}
x^3 & 1 & 2 \\[6pt]
x^2 & 1 & 8 \\[6pt]
x^1 & -6 & 0 \\[6pt]
x^0 & 8 &
\end{array}
$$

In the first column of the matrix, there are two changes of signs as $1 \to -6 \to 8$. Thus, there are two roots of $P(x)$ with positive real parts, and hence one root with negative real part. In fact, we can write $P(x) = (x + 2)(x^2 - x + 4)$, which has roots $x = -2$ and $x_{\pm} = 0.5 \pm i\sqrt{15}/2$.

Example 3.12 (A model for a measles epidemic) Measles is a childhood disease that has yet to be eradicated, despite global vaccination programs [21]. Measles usually occurs in preschool-age children before they receive their vaccines. During infancy, a baby may have maternal antibodies from the mother which are effective for up to one year after birth. These antibodies are responsible for immunity in infants while their immune system is developing during the first year of life. Vaccines are not recommended until these antibodies have waned.

In the presence of vaccination, a simple mathematical model to describe the dynamics of measles in the population can be expressed by the following system of differential equations [11]:

$$
\begin{aligned}
\frac{dS}{dt} &= (1 - p)\Lambda - \beta SI - \mu S, \\
\frac{dV}{dt} &= p\Lambda - \mu V + \tau I, \\
\frac{dE}{dt} &= \beta SI - \mu E - \sigma E, \\
\frac{dI}{dt} &= \sigma E - \mu I - \tau I,
\end{aligned}
\tag{3.13}
$$

where Λ is the rate at which individuals are born, p is the fraction of infants who are vaccinated, β is the rate at which susceptible individuals become infected by those who are infectious, μ is the natural death rate, σ is the rate at which exposed individuals become infectious, and τ is the rate at which infected individuals are treated and recovered with permanent immunity. It is assumed that vaccinated individuals have permanent immunity to the disease. It is also assumed that all the model parameters and variables are nonnegative. Here we analyze the stability of the critical points of the model.

In the absence of infection ($I = 0$), the model has a critical point C_0 given by:

$$C_0 = \left(\frac{(1-p)\Lambda}{\mu}, \frac{p\Lambda}{\mu}, 0, 0 \right).$$

This critical point is obtained by setting the right-hand side of the equations in model (3.13) to zero and solving for the variables while $I = 0$. To show the stability of C_0, we obtain the eigenvalues of the Jacobian matrix of the linearized system. The Jacobian at C_0 has the form:

$$J_0 = \begin{bmatrix} -\mu & 0 & 0 & -\frac{(1-p)\beta\Lambda}{\mu} \\ 0 & -\mu & 0 & \tau \\ 0 & 0 & -(\mu+\sigma) & \frac{(1-p)\beta\Lambda}{\mu} \\ 0 & 0 & \sigma & -(\mu+\tau) \end{bmatrix}.$$

The eigenvalues of this matrix are $\lambda_1 = \lambda_2 = -\mu$ and the roots of the quadratic equation:

$$P(\lambda) = \mu\lambda^2 + [2\mu^2 + \mu(\sigma+\tau)]\lambda + \mu(\mu+\sigma)(\mu+\tau) - \sigma\beta\Lambda(1-p).$$

The roots of P have negative real parts if and only if:

$$\mu(\mu+\sigma)(\mu+\tau) - \sigma\beta\Lambda(1-p) > 0.$$

Let us define the basic reproduction number [6]:

$$R_0 = \frac{\sigma\beta\Lambda(1-p)}{\mu(\mu+\sigma)(\mu+\tau)},$$

as discussed in Chapter 2. Thus, C_0 is locally asymptotically stable node if $R_0 < 1$ and unstable if $R_0 > 1$.

When $I \neq 0$, the model has another critical point, $C^* = (S^*, V^*, E^*, I^*)$, where

$$S^* = \frac{(\mu+\sigma)(\mu+\tau)}{\sigma\beta},$$

$$V^* = \frac{p\Lambda}{\mu} + \frac{\tau}{\beta}(R_0 - 1),$$

$$E^* = \frac{\mu(\mu+\tau)}{\sigma\beta}(R_0 - 1),$$

$$I^* = \frac{\mu}{\beta}(R_0 - 1).$$

Clearly, this critical point is positive if $R_0 > 1$. Otherwise, C_0 is the only positive critical point of the model. In order to determine the local stability of C^*, we define:

$$\alpha^* = \frac{\mu(\mu+\sigma)(\mu+\tau)}{\sigma\beta\Lambda}.$$

Thus,

$$\begin{cases} R_0 < 1 \text{ if } p > 1 - \alpha^*, \\ R_0 > 1 \text{ if } p < 1 - \alpha^*. \end{cases}$$

The Jacobian of (3.13) evaluated at C^* is:

$$J^* = \begin{bmatrix} -\mu R_0 & 0 & 0 & -\frac{(\mu+\sigma)(\mu+\tau)}{\sigma} \\ 0 & -\mu & 0 & \tau \\ \mu(R_0-1) & 0 & -(\mu+\sigma) & \frac{(\mu+\sigma)(\mu+\tau)}{\sigma} \\ 0 & 0 & \sigma & -(\mu+\tau) \end{bmatrix}.$$

The eigenvalues of J^* are $\eta_1 = -\mu$ and the roots of the polynomial:

$$Q(\eta) = \eta^3 + a\eta^2 + b\eta + c,$$

where

$$a = \mu R_0 + 2\mu + \sigma + \tau,$$
$$b = \mu R_0(2\mu + \sigma + \tau),$$
$$c = \mu(R_0 - 1)(\mu + \sigma)(\mu + \tau).$$

If $p < 1 - \alpha^*$, then $c > 0$. Since $b > 0$, it remains to show that $ab - c > 0$. Some algebraic calculations show that, since $\mu R_0 > \mu$, we have:

$$ab - c = \mu R_0(2\mu + \sigma + \tau)[\mu R_0 + (2\mu + \sigma + \tau)]$$
$$- \mu R_0(\mu + \sigma)(\mu + \tau) + \mu(\mu + \sigma)(\mu + \tau) > 0.$$

Thus, the Routh–Hurwitz criterion shows that the roots of Q have negative real parts whenever $p < 1 - \alpha^*$ ($R_0 > 1$), and therefore C^* is a stable node.

Exercises

1 Explore the dynamical behavior of the following model for population growth:

$$x' = xe^{r(1-x/N)},$$

where r and N are real constants.

2 Find an explicit solution of the following model of population growth and discuss its dynamical behavior:

$$x' = rx\left[1 - \left(\frac{x}{K}\right)^a\right],$$

where all parameters are positive and $0 < a \leq 1$.

3 When resources are limited, the average size of individuals in a population may begin to shrink over time. A model that explains this change in size can be expressed by the following system of differential equations:

$$\frac{dP}{dt} = rP\left(1 - \frac{PS}{K}\right),$$

$$\frac{dS}{dt} = b\left(\frac{S}{L} - 1\right)\left(1 - \frac{S}{M}\right)\left(1 - \frac{PS}{N}\right),$$

where P is the population size and S is the average size of individuals in the population. The parameters r and b are the rates of growth of population and size of individuals, respectively, and L represents the minimum size of individuals in the population. The maximum size of individuals in the population is denoted by M. We assume that if PS exceeds N, then the average size of the individuals begins to shrink. Describe the properties of this model, determine its critical points, and analyze their stability.

4 Find all the critical points of the following systems and analyze their stability:

a)

$$x' = x - y,$$
$$y' = x^2 + y^2 - 2.$$

b)

$$x' = x^2 - 3xy + 2x,$$
$$y' = x + y - 1.$$

c)

$$x' = x^2 - y - 1,$$
$$y' = (x - 2)y.$$

d)

$$x' = y - xy^2,$$
$$y' = yx^2 - x.$$

5 Determine the dimension of stable, unstable, and center manifolds of each of the following systems of differential equations around their critical points for $\mu \in \mathbb{R}$:

a)

$$x' = \mu - x + 2xy,$$
$$y' = (x + \mu)y.$$

b)

$$x' = x(1 - x) - \mu xy,$$
$$y' = (x - 1)y - yz,$$
$$z' = (y - 1)z.$$

c)

$$x' = x - y,$$
$$y' = (\mu - x)y - x^2.$$

6 Analyze the chemostat and Monod-type models of growth proposed in Exercises 4 and 5 of Chapter 2 for their critical points and the local behavior of solutions.

7 Consider the following system of differential equations:

$$x' = m - bxy - dx + az,$$
$$y' = bxy - (d + c)y,$$
$$z' = cy - (d + a)z,$$

where all the parameters are positive. Find the critical points and analyze their stability by proposing conditions on parameters.

8 Determine the number of roots with positive and negative real parts for the polynomial $P(x) = x^3 + 3ax^2 + (\alpha + 2)x + 4$, where $\alpha \in \mathbb{R}$.

9 Sepsis is a severe systemic inflammatory response syndrome caused by infection (e.g., bacteria) [46]. An overwhelming immune response is the main characteristic of this disease. Under normal conditions, the immune system can clear bacteria without causing inflammation. In some cases, however, a hyper-responsive immune system not only kills bacteria, but also damages healthy cells. If this situation occurs, patients may develop sepsis. It is suggested that the level of immune response in sepsis needs to be well regulated. For this purpose, a mathematical model can be useful to capture the kinetics of immune response during sepsis. A model describing the kinetics of interaction between bacteria (P) and immune (Kupffer) cells (K) can be expressed by the following system of differential equations:

$$\frac{dP}{dt} = r_p P \left(1 - \frac{P}{P_\infty}\right) - \frac{r_c P^n}{(P^n + k_c^n)} KP,$$

$$\frac{dK}{dt} = r_k K \left(1 - \frac{K}{K_\infty}\right) - \frac{r_c P^n}{(P^n + k_c^n)} KP - \delta_k K,$$

where $n = 0, 1, 2,$ and $r_p, P_\infty, r_c, k_c^n, r_k, K_\infty, \delta_k$ represent the rate of bacteria growth, bacteria carrying capacity, the rate at which bacteria are killed by Kupffer cells, the number of Kupffer cells which kill half of bacteria, the proliferation rate of Kupffer cells under inflammation, the maximum carrying capacity of Kupffer cells, and the killing rate of free Kupffer cells induced by binding to bacteria, respectively. Find the critical points of this system, analyze their stability, and sketch the associated phase plane.

10 Competition models are used to describe the dynamics of predators competing for the same or different resources. A model formulating such dynamics can be expressed by the following system of differential equations:

$$\frac{dP_1}{dt} = \alpha_1 P_1 \left(1 - \frac{P_1}{k_1} - \beta_{12} \frac{P_2}{k_1}\right),$$

$$\frac{dP_2}{dt} = \alpha_2 P_2 \left(1 - \frac{P_2}{k_2} - \beta_{21} \frac{P_1}{k_2}\right),$$

where P_1 and P_2 represent the population sizes of competing predators, α_1 and α_2 are the growth rate of P_1 and P_2 respectively, k_1 and k_2 are their carrying capacities, and β_{12} and β_{21} quantify the effect of each predator on the competing predator growth. Employ a suitable change of variables to nondimensionalize the model and obtain the following form:

$$\frac{dx}{dt} = x(1 - x - d_{12}y),$$

$$\frac{dy}{dt} = \phi y(1 - y - d_{21}x),$$

where $\phi, d_{12},$ and d_{21} are defined based on the parameters in new variables. Find the nullclines and show that there are four critical points. Analyze the local dynamics (stability) of the critical points, and draw the system phase plane.

4

Bifurcation

In this chapter, we consider some structural behavior of mathematical models expressed as systems of ordinary differential equations. A vector field (or phase plane) is structurally stable if for small changes (or perturbations) of the system parameters, qualitatively similar behavior is observed. If changes in a parameter of the system result in a structurally different phase plane (e.g., a change from stable to unstable solutions), then the system may undergo the phenomenon of *bifurcation*, and the parameter is called a *bifurcation parameter* [39].

Suppose the system $x' = f(x; \mu)$, where $x \in \mathbb{R}^n$ and $\mu \in \mathbb{R}$, has a nonhyperbolic critical point at $x = 0$ when $\mu = 0$. In this system, we consider μ as the bifurcation parameter. Using analysis tools discussed in Chapter 3 and center manifold theory [39], we investigate the behavior of the solutions near $x = 0$ when μ is near zero without analyzing the full n-dimensional system of equations. In a local neighborhood of the critical point $x = 0$, we can classify the type of bifurcation the system undergoes based on the eigenvalues of the Jacobian.

In this chapter, we restrict our attention to n-dimensional systems where $n = 1, 2$. For the case $n = 1$, expanding the system:

$$x' = f(x, \mu), \quad x, \mu \in \mathbb{R}, \tag{4.1}$$

as a Taylor series in x and μ gives:

$$x' = f(0,0) + f_x(0,0)x + f_\mu(0,0)\mu$$
$$+ \frac{1}{2}(f_{xx}(0,0)x^2 + 2f_{x\mu}(0,0)x\mu + f_{\mu\mu}(0,0)\mu^2) + \cdots$$

Since $x = 0$ is a nonhyperbolic critical point, we have $f(0,0) = f_x(0,0) = 0$, and the system dynamics around $(x, \mu) = (0,0)$ is determined by higher-order partial derivatives of f. Here, we consider three types of bifurcations, namely, the transcritical, saddle node, and pitchfork as detailed below.

Mathematical Modelling: A Graduate Textbook, First Edition. Seyed M. Moghadas and Majid Jaberi-Douraki.
© 2019 John Wiley & Sons, Inc. Published 2019 by John Wiley & Sons, Inc.
Companion Website: www.wiley.com/go/Moghadas/Mathematicalmodelling

4.1   Transcritical Bifurcation

In this type of bifurcation, at least one critical point of the system changes the stability at a critical value of the bifurcation parameter. To identify this bifurcation, we consider equation (4.1), with $f(0,0) = f_x(0,0) = 0$. The conditions for a transcrtical bifurcation can be summarized as follows (Figure 4.1).

Transcritical bifurcation [39]. *If $f_\mu(0,0) = 0$, $f_{xx}(0,0) \neq 0$, and $f_{\mu x}^2(0,0) - f_{xx}(0,0)f_{\mu\mu}(0,0) > 0$, then there are two continuous curves of critical points in some neighborhood of $(0,0)$ in the (x,μ) plane. These curves intersect transversally at $(0,0)$. For a sufficiently small $\mu \neq 0$, there are two hyperbolic critical points near $x = 0$, and:*
a) *the upper one is stable and the lower one is unstable if $f_{xx}(0,0) < 0$;*
b) *the upper one is unstable and the lower one is stable if $f_{xx}(0,0) > 0$.*

Example 4.1   Consider the following differential equation for x, $\mu \in \mathbb{R}$:

$$x' = \mu x - x^2.$$

It can be easily seen that the two critical points are $x = 0$ and $x = \mu$. For $x = 0$, the Jacobian $J = \mu - 2x$ gives $J_{(x=0)} = \mu$ and therefore:

$x = 0$ unstable if $\mu > 0$,
$x = 0$ stable if $\mu < 0$.

For $x = \mu$, the Jacobian gives $J_{(x=\mu)} = -\mu$, and therefore:

$x = \mu$ stable if $\mu > 0$,
$x = \mu$ unstable if $\mu < 0$.

We also find that $f_{xx}(0,0) = -2 < 0$, and therefore, for a sufficiently small μ, we can represent the bifurcation diagram by Figure 4.1(a). We can show that

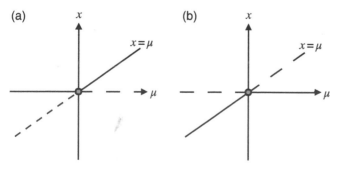

Figure 4.1 Schematic diagram of transcritical bifurcation with (a) $f_{xx}(0,0) < 0$ and (b) $f_{xx}(0,0) > 0$. Solid and dashed lines represent the presence of stable and unstable critical points, respectively.

the equation $x' = -\mu x + x^2$ undergoes a transcritical bifurcation at $(0,0)$, which corresponds to the bifurcation diagram shown in Figure 4.1(b).

Example 4.2 Consider the following differential equation for x, $\mu \in \mathbb{R}$:

$$x' = x(1 - x)(\mu - x).$$

The critical points are $x = 0$, $x = 1$, and $x = \mu$. The Jacobian $J = (1 - x)(\mu - x) - x(\mu - x) - x(1 - x)$ of the system gives $J_{(x=0)} = \mu$. Thus, $x = 0$ is stable if $\mu < 0$, and unstable if $\mu > 0$. Furthermore, $f_{xx} = -(\mu - x) - (1 - x) - (\mu - x) + x - (1 - x) + x = -2\mu + 6x - 2$, which is negative at $x = 0$ for a sufficiently small μ near zero. This implies that the system undergoes a transcritical bifurcation as μ passes through zero.

At the critical point $x = 1$, we get $J_{(x=1)} = 1 - \mu$, and thus $x = 1$ is stable if $\mu > 1$ and unstable if $\mu < 1$. Furthermore, $f_{xx} > 0$ at $x = 1$ for $\mu > 0$ and sufficiently close to 1, indicating a transcritical bifurcation.

When $x = \mu$, we get $J_{x=\mu} = -\mu(1 - \mu)$. In this case, $J_{x=\mu} > 0$ for $\mu < 0$, $J_{x=\mu} < 0$ for $0 < \mu < 1$, and $J_{x=\mu} > 0$ for $\mu > 1$. Thus, the nature of bifurcations is the same as above at $x = 0$ and $x = 1$.

4.2 Saddle-Node Bifurcation

A saddle-node bifurcation generally corresponds to the situation in which a saddle node and a stable node of the system approach each other and as the bifurcation parameter passes through its critical value, these critical points collide and disappear.

To identify this type of bifurcation, again we consider equation (4.1), with $f(0,0) = f_x(0,0) = 0$. The conditions for a saddle-node bifurcation can be summarized as follows.

Saddle-node bifurcation [39]. *If $f_\mu(0,0) \neq 0$ and $f_{xx}(0,0) \neq 0$, then there is a continuous curve of critical points in some neighborhood of $(0,0)$ in the (x, μ) plane such that the curve is tangent to the line $\mu = 0$, and:*

a) *if $f_\mu(0,0)f_{xx}(0,0) < 0$, then there are no critical points near $(0,0)$ for $\mu < 0$, while for any positive value of μ sufficiently close to zero, there are two critical points with x value near zero. These critical points are hyperbolic for $\mu \neq 0$, with the upper one being stable and the lower one unstable.*

b) *if $f_\mu(0,0)f_{xx}(0,0) > 0$, then there are no critical points near $(0,0)$ for $\mu > 0$, while for any negative value of μ sufficiently close to zero, there are two critical points with x value near zero. These critical points are hyperbolic for $\mu \neq 0$, with the upper one being unstable and the lower one stable.*

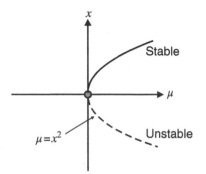

Figure 4.2 Schematic diagram of the saddle-node bifurcation in equation (4.2).

Example 4.3 Consider the following differential equation for x, $\mu \in \mathbb{R}$:

$$x' = \mu - x^2. \tag{4.2}$$

If $\mu < 0$, the system has no critical points. If $\mu > 0$, then there are two critical points, $x = \pm\sqrt{\mu}$. For $x = -\sqrt{\mu}$, the Jacobian $J = -2x$ gives $J_{(-\sqrt{\mu})} = 2\sqrt{\mu} > 0$, and therefore the critical point is unstable. For $x = \sqrt{\mu}$, the Jacobian is $J_{(\sqrt{\mu})} = -2\sqrt{\mu} < 0$, and hence the critical point is stable. Since $f_\mu(0,0)f_{xx}(0,0) = -2 < 0$, the upper branch is stable and the lower one is unstable (Figure 4.2).

Example 4.4 Consider the following differential equation for x, $\mu \in \mathbb{R}$:

$$x' = \mu + x^2. \tag{4.3}$$

If $\mu > 0$, the system has no critical points. If $\mu < 0$, then there are two critical points, $x = \pm\sqrt{-\mu}$. For $x = -\sqrt{-\mu}$, the Jacobian $J = 2x$ gives $J_{(-\sqrt{-\mu})} = -2\sqrt{-\mu} < 0$, and therefore the critical point is stable. For $x = \sqrt{-\mu}$, the Jacobian is $J_{(\sqrt{-\mu})} = 2\sqrt{-\mu} > 0$, and hence the critical point is unstable. Since $f_\mu(0,0)f_{xx}(0,0) = 2 > 0$, the upper branch is unstable and the lower one is stable (Figure 4.3).

4.3 Pitchfork Bifurcation

This type of bifurcation generally corresponds to the situation in which a stable node becomes unstable when the bifurcation parameter passes through a critical value, throwing off a pair of stable critical points of the system.

To investigate this type of bifurcation, let us consider equation (4.1), with $f(0,0) = f_x(0,0) = 0$. The conditions for a pitchfork bifurcation can be summarized as follows.

Figure 4.3 Schematic diagram of the saddle-node bifurcation in equation (4.3).

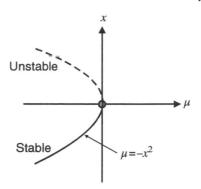

Pitchfork bifurcation [39]. *If* $f_\mu(0,0) = f_{xx}(0,0) = 0$, $f_{\mu x}(0,0) \neq 0$, *and* $f_{xxx}(0,0) \neq 0$, *then there are two continuous curves of critical points passing through* $(0,0)$, *where one of the curves is transverse to the axis* $\mu = 0$ *while the other curve is tangent to* $\mu = 0$ *at* $(0,0)$, *and:*

a) *if* $f_{\mu x}(0,0)f_{xxx}(0,0) < 0$, *then there exist three critical points near* $x = 0$ *for* $\mu > 0$ *and:*

   i) *the outer pair is stable and the inner one is unstable if* $f_{xxx}(0,0) < 0$. *There is only one critical point for* $\mu < 0$, *which is stable if* $f_{xxx}(0,0) < 0$.

   ii) *the outer pair is unstable and the inner one is stable if* $f_{xxx}(0,0) > 0$. *There is only one critical point for* $\mu < 0$, *which is unstable if* $f_{xxx}(0,0) > 0$.

b) *if* $f_{\mu x}(0,0)f_{xxx}(0,0) > 0$, *then there exist three critical points near* $x = 0$ *for* $\mu < 0$, *and:*

   i) *the outer pair is stable and the inner one is unstable if* $f_{xxx}(0,0) < 0$. *There is only one critical point for* $\mu > 0$, *which is stable if* $f_{xxx}(0,0) < 0$.

   ii) *the outer pair is unstable and the inner one is stable if* $f_{xxx}(0,0) > 0$. *There is only one critical point for* $\mu > 0$, *which is unstable if* $f_{xxx}(0,0) > 0$.

Example 4.5 Consider the following differential equation for x, $\mu \in \mathbb{R}$:

$$x' = \mu x - x^3. \tag{4.4}$$

The system always has critical point $x = 0$. If $\mu > 0$, then there are two additional critical points, given by $x = \pm\sqrt{\mu}$. The Jacobian of the system $J = \mu - 3x^2$ gives $J_{(x=0)} = \mu$, and therefore $x = 0$ is stable if $\mu < 0$ and unstable if $\mu > 0$.

If $\mu > 0$, $J_{(\pm\sqrt{\mu})} = -2\mu < 0$, and therefore both critical points $x = \pm\sqrt{\mu}$ are stable. It can be seen that $f_{\mu x}(0,0)f_{xxx}(0,0) = -6 < 0$, and $f_{xxx}(0,0) = -6 < 0$, corresponding to the bifurcation diagram in Figure 4.4.

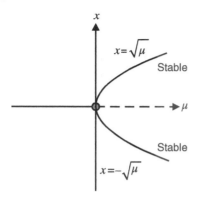

Figure 4.4 Schematic diagram of the pitchfork bifurcation in equation (4.4).

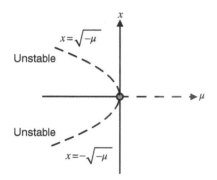

Figure 4.5 Schematic diagram of the pitchfork bifurcation in equation (4.5).

Example 4.6 Consider the following differential equation for x, $\mu \in \mathbb{R}$:

$$x' = \mu x + x^3. \tag{4.5}$$

This system has critical point $x = 0$. If $\mu < 0$, then there are two additional critical points, given by $x = \pm\sqrt{-\mu}$. The Jacobian at $x = 0$ is $J_0 = \mu$, and therefore $x = 0$ is stable if $\mu < 0$ and unstable if $\mu > 0$.

If $\mu < 0$, $J_{(\pm\sqrt{-\mu})} = -2\mu > 0$, and therefore both critical points $x = \pm\sqrt{\mu}$ are unstable. We also note that $f_{\mu x}(0,0)f_{xxx}(0,0) = 6 > 0$, and $f_{xxx}(0,0) = 6 > 0$, which leads to a pitchfork bifurcation as shown in Figure 4.5.

4.4 Hopf Bifurcation

The bifurcations discussed in the previous section can occur in one-dimensional systems. In this section, we consider another type of bifurcation that requires a higher-dimensional space to occur. We restrict our attention to a two-dimensional space ($\mathbb{R}^2$).

A Hopf bifurcation refers to a structural change in the system behavior around a critical point, say X^*, and corresponds to the scenario in which the

Jacobian at X^* exhibits a pair of purely imaginary eigenvalues (with zero real part) when the bifurcation parameter passes through a critical value [39, 45].

To identify this type of bifurcation, consider the following two-dimensional system of differential equations:

$$x' = f(x, y, \mu),$$
$$y' = g(x, y, \mu), \tag{4.6}$$

where $(x, y) \in \mathbb{R}^2$ and $\mu \in \mathbb{R}$. Suppose $X^* = (x^*, y^*)$ is a critical point, and the linearized system around X^* has the eigenvalues:

$$\lambda_\pm(\mu) = \alpha(\mu) \pm \beta(\mu)i,$$

as a function of the bifurcation parameter μ. Suppose further that at a critical value $\mu = \mu^*$, the following conditions are satisfied:

1) $\alpha(\mu^*) = 0$ and $\beta(\mu^*) = \omega \neq 0$, where

$$\text{sign}(\omega) = \text{sign}\left(\frac{\partial g}{\partial x}\right)\Big|_{(x^*, y^*, \mu^*)}.$$

We refer to this as the *nonhyperbolic* condition.

2) $\left(\frac{d\alpha(\mu)}{d\mu}\right)\Big|_{\mu^*} = d \neq 0$. This condition is referred to as the *transversality* condition.

3) $a \neq 0$, where

$$a = \frac{1}{16}(f_{xxx} + f_{xyy} + g_{xxy} + g_{yyy})|_{(x^*, y^*, \mu^*)}$$
$$+ \frac{1}{16\omega}[f_{xy}(f_{xx} + f_{yy}) - g_{xy}(g_{xx} + g_{yy}) - f_{xx}g_{xx} + f_{yy}g_{yy}]|_{(x^*, y^*, \mu^*)}.$$

This is called the *genericity* condition.

Hopf bifurcation [39]. *If the above conditions are satisfied, then a unique curve of periodic solutions bifurcates from X^* into the region $\mu > \mu^*$ if $ad < 0$ and into the region $\mu < \mu^*$ if $ad > 0$. In this case,*
   a) *the critical point X^* is stable for $\mu > \mu^*$ and unstable for $\mu < \mu^*$ if $d < 0$;*
   b) *the critical point X^* is stable for $\mu < \mu^*$ and unstable for $\mu > \mu^*$ if $d > 0$.*

The periodic solutions are generally stable if X^* is unstable on the side of $\mu = \mu^*$ where the periodic solutions exist. Similarly, the periodic solutions are generally unstable if X^* is stable on the side of $\mu = \mu^*$ where the periodic solutions exist. The Hopf bifurcation is called *supercritical* if periodic solutions are stable and *subcritical* if the periodic solutions are unstable.

Example 4.7   Consider the following differential equation for $x, \mu \in \mathbb{R}$:

$$x'' - (\mu - x^2)x' + x = 0.$$

Using the change of variables $u = x$ and $v = x'$, the equation transforms into the following system of differential equations:

$$u' = v,$$
$$v' = -u + (\mu - u^2)v. \tag{4.7}$$

The system has a unique critical point at $(0,0)$. The Jacobian of the system at this critical point is:

$$J_{(0,0)} = \begin{bmatrix} 0 & 1 \\ -1 - 2uv & \mu - u^2 \end{bmatrix}_{(0,0)} = \begin{bmatrix} 0 & 1 \\ -1 & \mu \end{bmatrix}.$$

Solving the characteristic equation yields the eigenvalues $\lambda = \alpha(\mu) \pm \beta(\mu)i$ for sufficiently small μ near zero, where

$$\alpha(\mu) = \frac{1}{2}\mu,$$

$$\beta(\mu) = \frac{\sqrt{4 - \mu^2}}{2}.$$

At $\mu = 0$, we have $\alpha(0) = 0$, and therefore the system has purely imaginary eigenvalues. Furthermore,

$$d = \left(\frac{d\alpha}{d\mu}\right)\Big|_{\mu=0} = \frac{1}{2} \neq 0,$$

$$\left(\frac{\partial g}{\partial u}\right)\Big|_{(0,0,0)} = (-1 - 2uv)|_{(0,0,0)} = -1,$$

and therefore $\omega = -1$. Calculating a (in the genericity condition of a Hopf bifurcation) gives $a = -\frac{1}{8}$ and hence $ad = -\frac{1}{16} < 0$. Thus, the Hopf bifurcation is supercritical and $(0,0)$ is stable for $\mu < 0$ and unstable for $\mu > 0$. The presence of a periodic solution in the region $\mu > 0$ as a result of the Hopf bifurcation is shown in Figure 4.6.

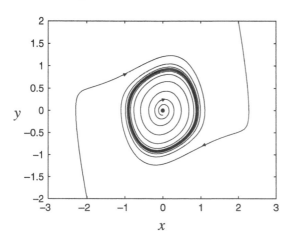

Figure 4.6 Phase plane of system (4.7) for the presence of a period solution when the system undergoes a Hopf bifurcation with $\mu = 0.2$. The period solution is stable and the Hopf bifurcation is supercritical.

Example 4.8 Consider the following system of differential equations for x, y, $\mu \in \mathbb{R}$:

$$x' = y,$$
$$y' = -x + \mu y - y^3.$$

The Jacobian of the system at the critical point $(x, y) = (0, 0)$ is:

$$J_{(0,0)} = \begin{bmatrix} 0 & 1 \\ -1 & \mu \end{bmatrix},$$

with characteristic equation $\lambda^2 - \mu\lambda + 1 = 0$. A similar argument to Example 4.7 can be used to determine the type of Hopf bifurcation the system undergoes, and we leave this as an exercise.

For time-dependent systems, a periodic solution arising from a Hopf bifurcation can also be represented by oscillatory behavior of the system variables in time. For example, in a Gause-type predator–prey model, given by:

$$x' = rx\left(1 - \frac{x}{K}\right) - \frac{axy}{1 + x},$$
$$y' = \left(\frac{cx}{1 + x} - d\right)y, \tag{4.8}$$

the system may undergo a Hopf bifurcation for a range of parameter values, which leads to the appearance of a period solution in the (x, y) plane. This solution can be represented by oscillatory dynamics of x and y as functions of time, which reflect the variations in population sizes of preys and predators in an ecosystem. Figure 4.7 shows such behavior for the model variables.

Example 4.9 Consider the following system of differential equations:

$$\frac{dx}{dt} = -2x + \mu y + x^2 y \equiv f(x, y),$$
$$\frac{dy}{dt} = k - \mu y - x^2 y \equiv g(x, y), \tag{4.9}$$

where k and μ are real parameters in $\mathbb{R}$. We now discuss the bifurcation behavior of the system at its critical points. Solving the right-hand side of the equations gives a critical point of the system:

$$E^* = \left(\frac{k}{2}, \frac{4k}{4\mu + k^2}\right).$$

If $\mu \geq 0$, then E^* is the only critical point. If $\mu < 0$, then the system has no critical points. We restrict our attention to the case where $k, \mu > 0$. The Jacobian at E^* is:

$$J_{E^*} = \begin{bmatrix} -2 + \dfrac{4k^2}{4\mu + k^2} & \mu + \dfrac{k^2}{4} \\ -\dfrac{4k^2}{4\mu + k^2} & -\mu - \dfrac{k^2}{4} \end{bmatrix},$$

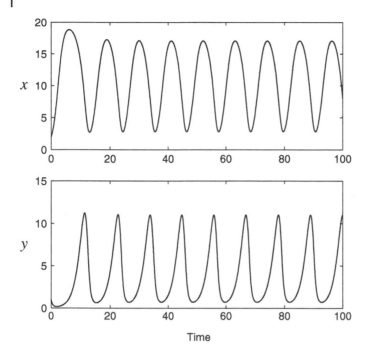

Figure 4.7 Oscillatory dynamics of prey (x) and predator (y) for model (4.8) over time with $r = 1, a = 1, c = 9, d = 8$, and $K = 20$.

with characteristic equation:

$$\lambda^2 - \left(\frac{2k^2 - 8\mu}{4\mu + k^2} - \mu - \frac{k^2}{4} \right) \lambda + 2 \left(\mu + \frac{k^2}{4} \right) = 0.$$

Assuming k as the bifurcation parameter, the eigenvalues are:

$$\lambda_{\pm} = \alpha(k) \pm i\beta(k),$$

where

$$\alpha(k) = -1 + \frac{2k^2}{4\mu + k^2} - \frac{4\mu + k^2}{8},$$

$$\beta(k) = \pm \frac{\sqrt{\left(\frac{2k^2 - 8\mu}{4\mu + k^2} - \mu - \frac{k^2}{4} \right)^2 - 8 \left(\mu + \frac{k^2}{4} \right)}}{2}.$$

For the system to satisfy the conditions of a Hopf bifurcation, we need to have the following relationships between k and μ, which result from solving $\alpha(k) = 0$ in terms of k:

$$k_1(\mu) = 2\sqrt{1 - \mu + \sqrt{1 - 4\mu}},\qquad\qquad(4.10)$$

or

$$k_2(\mu) = 2\sqrt{1 - \mu - \sqrt{1 - 4\mu}}. \tag{4.11}$$

A necessary condition for k to be a real number (and positive) is $\mu < 1/4$. The derivative of $\alpha(k)$ evaluated at k_1 and k_2 is:

$$\left.\frac{d\alpha}{dk}\right|_{k_1} = \sqrt{1 - \mu + \sqrt{1 - 4\mu}}\left(\frac{2\mu}{(1 + \sqrt{1 - 4\mu})^2} - \frac{1}{2}\right),$$

and

$$\left.\frac{d\alpha}{dk}\right|_{k_2} = \sqrt{1 - \mu - \sqrt{1 - 4\mu}}\left(\frac{2\mu}{(1 - \sqrt{1 - 4\mu})^2} - \frac{1}{2}\right).$$

Simple calculations yield that:

$$\left.\frac{d\alpha}{dk}\right|_{k_1} = 0, \quad \text{if } \mu_1 = \frac{1}{4},$$

and

$$\left.\frac{d\alpha}{dk}\right|_{k_2} = 0, \quad \text{if } \mu_2 = \frac{1}{4}.$$

We also note that:

$$\left.\frac{dg(x, y)}{dx}\right|_{E^*,k1} < 0, \quad \left.\frac{dg(x, y)}{dx}\right|_{E^*,k2} < 0.$$

Thus, from the genericity condition for Hopf bifurcation (left as an exercise), we find that the system undergoes two Hopf bifurcations when k passes through $k_1(\mu_1)$ and $k_2(\mu_2)$. Figure 4.8 shows the periodic solution arising (at k_1) and disappearing (at k_2) when $\mu = 0.1$.

4.5 Solution Types

The bifurcations discussed in this chapter often lead to different types of solutions. Here, we define several types of solutions commonly analyzed in dynamical systems theory [39].

Limit cycle. A limit cycle is a simple (i.e., it does not cross itself) closed solution of the system that contains no critical points and is periodic with respect to time. The periodic nature of a limit cycle indicates that every point moving along the curve will return to its starting point at fixed time intervals. This type of solution often arises when the system undergoes a Hopf bifurcation. The difference between a limit cycle and any other closed solution (with periodic behavior) is that the limit cycle represents the limiting behavior of the solutions in a neighborhood of any point on its curve.

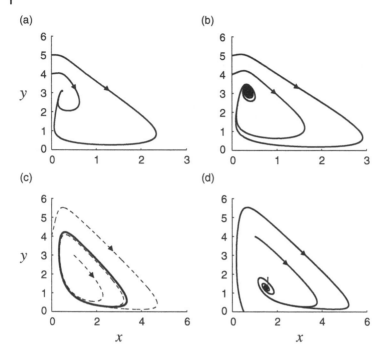

Figure 4.8 Phase-plane behavior of solutions with $\mu = 0.1$ and (a) $k = 0.5$ (stable critical point); (b) $k = 0.7$ (stable critical point); (c) $k = 2$ (stable periodic solution and unstable critical point); (d) $k = 3$ (stable critical point).

Stable limit cycle. A limit cycle L is stable if every solution $X(t)$ starting in a small neighborhood of L has the limit $\lim_{t\to+\infty} X(t) = L$ (Figure 4.9(a)).

Unstable limit cycle. A limit cycle L is unstable if every solution $X(t)$ starting in a small neighborhood of L has the limit $\lim_{t\to-\infty} X(t) = L$ (Figure 4.9(b)).

Semi-stable limit cycle. A limit cycle L is semi-stable if one of the following conditions is satisfied:

a) For every solution $X(t)$ starting in a small neighborhood of L inside L and every solution $Y(t)$ starting in a small neighborhood of L outside L (Figure 4.10(a)),

$$\lim_{t\to-\infty} X(t) = L, \quad \lim_{t\to+\infty} Y(t) = L.$$

b) For every solution $X(t)$ starting in a small neighborhood of L inside L and every solution $Y(t)$ starting in a small neighborhood of L outside L (Figure 4.10(b)),

$$\lim_{t\to+\infty} X(t) = L, \quad \lim_{t\to-\infty} Y(t) = L.$$

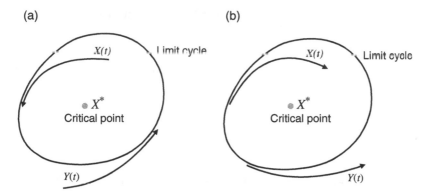

Figure 4.9 Representation of (a) stable and (b) unstable limit cycles.

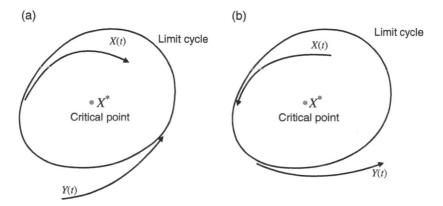

Figure 4.10 Representation of semi-stable limit cycles: (a) stable from outside and unstable from inside; (b) stable from inside and unstable from outside.

Homoclinic orbit. This represents a type of solution, $X(t)$, with the following properties:

$$\lim_{t \to +\infty} X(t) = \lim_{t \to -\infty} X(t) = X^*,$$

where X^* is a critical point of the system (Figure 4.11). This type of solution often arises when a saddle-node bifurcation and a Hopf bifurcation merge.

Heteroclinic orbit. This represents a type of solution, $X(t)$, with the following properties:

$$\lim_{t \to -\infty} X(t) = X_1^* \quad \text{and} \quad \lim_{t \to +\infty} X(t) = X_2^*,$$

where X_1^* and X_2^* are two distinct isolated critical points of the system (Figure 4.12).

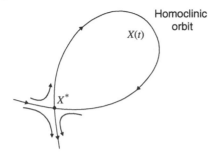

Homoclinic orbit

Figure 4.11 Representation of a homoclinic orbit.

$X(t)$

X^*

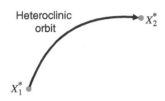

Heteroclinic orbit

X_2^*

Figure 4.12 Representation of a heteroclinic orbit.

X_1^*

Example 4.10 Consider the following system of differential equations:

$$x' = -y + x(\mu - x^2 - y^2),$$
$$y' = x + y(\mu - x^2 - y^2),$$

(4.12)

where $\mu \in \mathbb{R}$. First, we show that the system undergoes a Hopf bifurcation when μ passes through 0. Then we show that, for $\mu > 0$, a limit cycle appears (as a result of the Hopf bifurcation), which is stable.

The system has a unique critical point at $(0, 0)$. The Jacobian at this critical point is:

$$J_{(0,0)} = \begin{bmatrix} \mu & -1 \\ 1 & \mu \end{bmatrix}.$$

The characteristic equation is $\lambda^2 - 2\mu\lambda + \mu^2 + 1 = 0$, which has roots (i.e., the eigenvalues of $J_{(0,0)}$):

$$\lambda_{\pm}(\mu) = \frac{2\mu \pm \sqrt{4\mu^2 - 4(1 + \mu^2)}}{2} = \mu \pm i.$$

One can easily check the conditions for a Hopf bifurcation at $\mu = 0$. Here we focus on the stability of the limit cycle arising from this bifurcation for $\mu > 0$. Let $x = r\cos\theta$ and $y = r\sin\theta$ in polar coordinates. Thus, $x^2 + y^2 = r^2$ and $\tan\theta = y/x$. Taking the derivative gives:

$$rr' = xx' + yy' = -xy + x^2(\mu x - y + xy^2) + xy + y^2(\mu - x^2 - y^2)$$
$$= r^2(\mu - r^2).$$

Figure 4.13 The stable limit cycle for system (4.12) with $\mu = 2$.

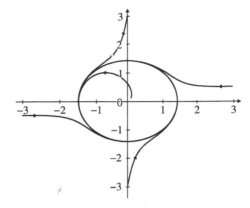

A simple calculation shows that $r^2\theta' = xy' - yx' = r^2$. Therefore, we can express the system in polar coordinates as:

$$r' = r(\mu - r^2),$$
$$\theta' = 1.$$

When $\mu < 0$, we have $r' < 0$, and therefore the solutions approach the origin $r = 0$, indicating that $(0, 0)$ in the original system is a stable node. When $\mu > 0$, we have:

$$r' > 0 \quad \text{if } 0 < r < \sqrt{\mu},$$
$$r' < 0 \quad \text{if } r > \sqrt{\mu}.$$

This shows that the limit cycle $r = \sqrt{\mu}$ is stable. Figure 4.13 shows the stable limit cycle for $\mu = 2$.

Exercises

1 Investigate the bifurcation behavior of $x' = f(x, \mu)$, where x, $\mu \in \mathbb{R}$ and:
   a) $f(x, \mu) = \mu^2 x + x^3$,
   b) $f(x, \mu) = x - \mu x(1 - x)$,
   c) $f(x, \mu) = x(\mu - e^x)$,
   d) $f(x, \mu) = \mu x - \ln(1 + x)$,
   e) $f(x, \mu) = \mu + \frac{x}{2} - \frac{x}{1+x}$,
   f) $f(x, \mu) = \mu + x - \ln(1 + x)$,
   g) $f(x, \mu) = x + \frac{\mu x}{1+x^2}$,
   h) $f(x, \mu) = \mu x - \frac{x}{1+x^2}$,
   i) $f(x, \mu) = \mu x + \frac{x^3}{1+x^2}$,
   j) $f(x, \mu) = \mu^2 ax + 2\mu ax^3 - x^5$, $\quad \alpha \in \mathbb{R}$.

2 For $x, \mu \in \mathbb{R}$, consider the following differential equation:

$$x' = \mu x - 2x^2 + x^3.$$

a) Determine the stability of the critical point $x = 0$. Identify the value of the bifurcation parameter μ and determine the type of bifurcation that occurs at this value. Sketch the bifurcation diagram.

b) Investigate whether the equation can exhibit additional critical points. Determine the conditions for the existence of these critical points and identify the type of bifurcations that can take place.

3 For each of the following systems of differential equations, a Hopf bifurcation occurs at the origin when $\mu = 0$. Investigate the type of Hopf bifurcation the systems undergo.

a)

$$x' = y + \mu x,$$
$$y' = -x + \mu y - x^2 y.$$

b)

$$x' = y + \mu x - x^2,$$
$$y' = -x + \mu y - x^2 y + 2x^2.$$

4 Using the change of variables $r^2 = x^2 + y^2$ and $\tan\ \theta = y/x$, show that the following system of differential equations has a stable limit cycle that contains the origin and sketch the phase plane:

$$x' = y + x(1 - x^2 - y^2),$$
$$y' = -x + y(1 - x^2 - y^2).$$

5 Using the same change of variables in the previous exercise, show that the following system of differential equations has an unstable limit cycle containing the origin and sketch the phase plane:

$$x' = y,$$
$$y' = x^2 y - y - x.$$

6 Consider the following system of differential equations:

$$r' = \alpha(\beta - r^3)r, \qquad \theta' = \omega,$$

where α and β are parameters. Explain the dynamics of this system by changing the parameter β. This system can also be written in (x, y) coordinates using the change of variables $r^2 = x^2 + y^2$ and $\tan\ \theta = y/x$.

7 Determine the type of bifurcation that the following systems may undergo, where x, y, and μ are real values:

a)

$$x' = x(x^2 + \mu - 1).$$

b)

$$x' = \mu x + y - x^2,$$
$$y' = -x + \mu y + 2x^2.$$

5

Discretization and Fixed-Point Analysis

The level of complexity in many mathematical models precludes the formulation of their explicit solutions. Therefore, understanding the qualitative dynamics of such models may require the study of their quantitative behavior [2, 50]. These models often depend on parameters that play crucial roles in system dynamics such as stability and bifurcation. To observe such dynamics, discretization and simulation tools are widely used to approximate the solutions of continuous models in the domain of their variables [4, 30]. When discretization is applied to a continuous model, we obtain a *discrete system* that can be represented by a set of difference equations. A general form of a discrete system with the independent variable of time is given by:

$$X(t+1) = F(t, X(t)),$$

where

$$X(t) = \begin{pmatrix} x_1(t) \\ x_2(t) \\ \vdots \\ x_n(t) \end{pmatrix}, \quad F(t, X(t)) = \begin{pmatrix} f_1(t, x_1(t), x_2(t), \ldots, x_n(t)) \\ f_2(t, x_1(t), x_2(t), \ldots, x_n(t)) \\ \vdots \\ f_n(t, x_1(t), x_2(t), \ldots, x_n(t)) \end{pmatrix}, \quad (5.1)$$

and f_i is a continuous real-valued function. In this system, the vector $X(t)$ can be calculated iteratively. Suppose, for $t = t_0$, the value of $X(t_0)$ is known. Substituting this vector into $F(t_0, x_1(t_0), x_2(t_0), \ldots, x_n(t_0))$ gives the vector $X(t_0 + 1)$. We can continue this process and calculate $X(t_0 + n)$ for $n \geq 1$. This iteration method can be generalized to $X(t + h) = F(t, X(t))$ for any fixed $h > 0$.

When the solutions of a continuous system cannot be written in closed form, we may construct a discrete system to approximate their values using an initial value for the solution vector. Suppose $X'(t) = F(t, X(t))$ is a continuous system with $X(t_0) = X_0$. Let us consider the following difference equation:

$$X(t+h) = X(t) + hF(t, X(t)),$$

where $h > 0$ is a fixed number. Rearranging this equation and taking the limit when $h \to 0$ gives the continuous system. This shows that as h decreases, the

Mathematical Modelling: A Graduate Textbook, First Edition. Seyed M. Moghadas and Majid Jaberi-Douraki.
© 2019 John Wiley & Sons, Inc. Published 2019 by John Wiley & Sons, Inc.
Companion Website: www.wiley.com/go/Moghadas/Mathematicalmodelling

vector X in the difference equation with $X(t_0) = X_0$ provides a closer approximation to the solution of the continuous system.

To approximate the solution of a continuous system, the first step is to determine a corresponding discrete system by a set of difference equations. The formulation of a discrete system depends on the method used for discretization of the continuous system. Here we discuss two discretization schemes for deterministic models, including (i) the Euler method, and (ii) a nonstandard finite-difference method. The use of these methods for computer simulations will be discussed in Chapter 8.

5.1 Discretization

The process of developing a system of difference equations corresponding to a continuous time-dependent system is called *discretization*. In this process, the solutions of the continuous system are approximated by a finite number of points in a time interval, which depend on a parameter called the *time-step* [30]. This parameter characterizes the discretization process, and has no counterpart in the continuous system [2]. As the time-step becomes smaller (i.e., approaches 0) the number of points in the time interval increases, which is generally expected to improve the approximations made by the discrete system. We begin by describing the standard method of Euler for discretizing time-dependent variables in deterministic models. It is worth mentioning that the class of standard methods includes a number of well-established discretization processes such as Runge–Kutta methods [16]. Here, we restrict our attention to Euler's first-order method.

5.1.1 Euler Method

Suppose that $x' = f(t, x(t))$ is defined for $t \in [a, b]$ with the initial value $x(a) = x_0$. The Euler method discretizes the model on equal sub-intervals of t in $[a, b]$. Let $N > 0$ be an integer, and define $h = (b - a)/N$. Thus, the interval $[a, b]$ can be divided into N sub-intervals, where $h = t_{i+1} - t_i$ (Figure 5.1). In this process, t_i and t_{i+1} are referred to as the current and advanced time-steps, respectively. Using the Taylor expansion of the model at t_{i+1}, we have:

$$x(t_{i+1}) = x(t_i) + (t_{i+1} - t_i)x'(t_i) + \frac{(t_{i+1} - t_i)^2}{2!}x''(\xi),$$

where $\xi \in (t_i, t_{i+1})$. Ignoring the second-order terms provides an approximation in the form of a difference equation associated with the Euler method, given by:

$$x(t_{i+1}) = x(t_i) + hf(t_i, x(t_i)). \tag{5.2}$$

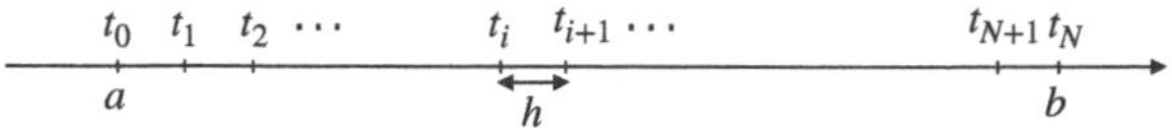

Figure 5.1 Discretization of the interval $[a, b]$ with a fixed time-step h.

When $h \to 0$, from (5.2) we obtain the continuous model. Since (5.2) holds as an approximation for every point of t in $[a, b]$, we have the general form:

$$x(t + h) = x(t) + hf(t, x(t)),$$

which is a *difference equation* for fixed $h > 0$. In this method, approximation begins by providing an initial condition $x(t_0) = x_0$ for the solution. Then, the method approximates $x(t_1)$ using (5.2), where $t_1 = t_0 + h$. This process continues iteratively to approximate the values of x at all t_i in the interval $[a, b]$. Connecting the approximated values of x represents the numerical trajectory of the solution x with the initial condition x_0.

Example 5.1   Consider the predator–prey system:

$$x' = rx(1 - x) - axy \equiv f_1(t, x(t), y(t)),$$
$$y' = (cx - d)y \equiv f_2(t, x(t), y(t)),$$

(5.3)

in which x and y are the population sizes for prey and predator, respectively, and the parameters $r, a, c,$ and d are positive. Discretizing the model using the Euler method gives:

$$x(t_{i+1}) = x(t_i) + h[rx(t_i)(1 - x(t_i)) - ax(t_i)y(t_i)],$$
$$y(t_{i+1}) = y(t_i) + h[cx(t_i) - d]y(t_i).$$

(5.4)

If the initial conditions $x(t_0)$ and $y(t_0)$ are known, then the solutions $x(t)$ and $y(t)$ can be approximated for a finite number of t_i.

As mentioned before, the accuracy of the method in approximating the system solutions improves as the time-step decreases. To show this numerically, we consider (5.4) with $r = 1$, $a = 0.9$, $c = 0.3$, and $d = 0.8$. Figure 5.2 shows numerical calculations with four different time-steps for the solution starting at $(1, 1.5)$. As h decreases, approximations are made closer to the solution curve.

5.1.2 Nonstandard Methods

It has been shown that standard methods such as Euler and Runge–Kutta may fail to preserve some fundamental properties of the underlying continuous model, especially in biological systems where the positivity of solutions may be a condition for system variables [33]. For example, it is entirely possible to have a negative approximation of $x(t_{i+1})$ in (5.2) if $f(t_i, x(t_i)) < 0$, even when $x(t_i) > 0$.

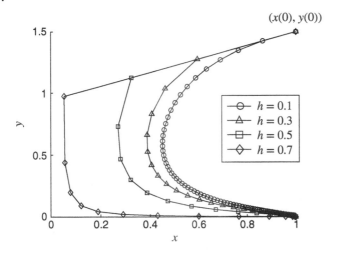

$(x(0), y(0))$

Legend:
- $h = 0.1$
- $h = 0.3$
- $h = 0.5$
- $h = 0.7$

Figure 5.2 Approximations using the Euler method for the solution of (5.3) with $h = 0.1$, 0.3, 0.5, 0.7.

Previous studies involving numerical approximation of deterministic models have reported the following issues for a number of standard methods [32, 33]:

i) The method may diverge for some values of the time-step.
ii) The use of a small time-step does not guarantee the convergence of the method to the correct solutions established by the qualitative dynamics of the system.
iii) The method may fail to be consistent with the transient behavior of the model.
iv) The method is affected by the initial conditions and may diverge or fail to capture the real dynamics of the model.

For these reasons, a number of nonstandard methods have been developed to ensure that the numerical solutions preserve the properties of the model [30]. Here, we introduce a simple first-order nonstandard finite-difference method with the associated rules for its construction.

☞ **Construction rules.** Suppose that $x' = f(t, x(t))$ is defined for $t \in [a, b]$ with the initial value $x(a) = x_0$. To develop the algorithm for the nonstandard method, we consider the same discretization of $[a, b]$ as in the Euler method with a fixed time-step h, and apply the following three "abc" rules.

a) All the terms of the same variable that is being approximated on the right-hand side of the equation $x' = f(t, x(t))$ with negative sign should be written with x in the advance time-step factored out (see below), that is, at time $t_{i+1} = t_i + h$. This means in practice that the negative terms for this

variable will move to the left-hand side of the equation for approximation (i.e., preserving positivity).
b) Terms in x with order higher than 1 on the right-hand side of $x' = f(t, x(t))$ with negative sign should be split into two parts:
   i) Order 1 should be approximated at the advance time-step $t_{i+1} = t_i + h$.
   ii) The remaining orders should be approximated at the current time-step t_i.
c) Every other variable (if it exists) on the right-hand side of the equation remains in the current time-step t_i.

In general, we can write the equation $x' = f(t, x(t))$ in the form:

$$x'(t) = U(t, x(t)) - x V(t, x(t)) \equiv f(t, x(t)),$$

where U and V are nonnegative real-valued functions [2]. Using the abc rules above, a form of difference equation is given by:

$$\frac{x(t_i + h) - x(t_i)}{h} = U(t_i, x(t_i)) - x(t_i + h) V(t_i, x(t_i)).$$

Rearranging this equation provides a nonstandard difference equation:

$$x(t_i + h) = \frac{x(t_i) + h U(t_i, x(t_i))}{1 + h V(t_i, x(t_i))}. \tag{5.5}$$

Example 5.2 Let us consider the logistic equation:

$$N'(t) = rN(t) \left(1 - \frac{N(t)}{K} \right) = rN(t) - \frac{r}{K} N^2(t),$$

where r, $K > 0$. We note that the second term on the right-hand side of the logistic equation has negative sign and is of order 2. Thus, we split this term according to the abc rules. Also the first term is positive and therefore is approximated by the value of N at the current time-step t_i. To discretize the model in a nonstandard form, we write the approximating equation in the form:

$$\frac{N(t_{i+1}) - N(t_i)}{h} = rN(t_i) - \frac{r}{K} N(t_{i+1}) N(t_i).$$

Rearranging for terms with the advanced time-step on the left-hand side of the equation, we get:

$$N(t_{i+1}) = \frac{K(1 + hr)N(t_i)}{K + hrN(t_i)}.$$

This approximation shows that the population size remains in the nonnegative domain if the initial condition is positive, regardless of the value of time-step h.

Example 5.3 Consider the predator–prey system in (5.3). In the previous section, we derived the Euler method for discretizing this system. Here, we

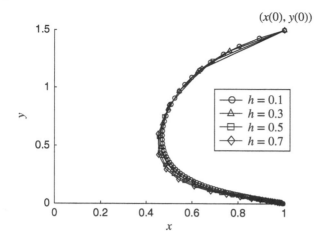

Figure 5.3 Approximations using the nonstandard method for the solution of (5.3) with $h = 0.1, 0.3, 0.5, 0.7$.

develop a nonstandard method for each variable. Based on the abc rules, we have:

$$\frac{x(t_{i+1}) - x(t_i)}{h} = rx(t_i) - rx(t_{i+1})x(t_i) - ax(t_{i+1})y(t_i),$$

$$\frac{y(t_{i+1}) - y(t_i)}{h} = cx(t_i)y(t_i) - dy(t_{i+1}). \tag{5.6}$$

Rearranging these equations, we obtain the method as:

$$x(t_{i+1}) = \frac{(1 + hr)x(t_i)}{1 + hrx(t_i) + hay(t_i)},$$

$$y(t_{i+1}) = \frac{(1 + hcx(t_i))y(t_i)}{1 + hd}. \tag{5.7}$$

For the purpose of comparison, let us perform numerical calculations with (5.7) with the same parameter values used in Example 5.1. Figure 5.3 shows the approximations with the same four different time-steps for the solution starting at $(1, 1.5)$. Clearly, the nonstandard method outperforms the Euler method in approximating the solution, especially when h increases.

Example 5.4 Suppose we have the following model that describes the spread of an epidemic in the population:

$$S' = -\beta SI + \delta R,$$

$$I' = \beta SI - \gamma I, \tag{5.8}$$

$$R' = \gamma I - \delta R.$$

In this model, the parameters β (transmission rate), γ (recovery rate), and δ (rate of loss of immunity) are positive. Using the first-order derivative for each variable and applying the abc rules, we can discretize the system as:

$$\frac{S(t_{i+1}) - S(t_i)}{h} = -\beta S(t_{i+1})I(t_i) + \delta R(t_i),$$

$$\frac{I(t_{i+1}) - I(t_i)}{h} = \beta S(t_i)I(t_i) - \gamma I(t_{i+1}),$$

$$\frac{R(t_{i+1}) - R(t_i)}{h} = \gamma I(t_i) - \delta R(t_{i+1}).$$

Rearranging these equation gives the nonstandard method:

$$S(t_{i+1}) = \frac{S(t_i) + h\delta R(t_i)}{1 + h\beta I(t_i)},$$

$$I(t_{i+1}) = \frac{(1 + h\beta S(t_i))I(t_i)}{1 + h\gamma},$$

$$R(t_{i+1}) = \frac{R(t_i) + h\gamma I(t_i)}{1 + h\delta}.$$

A set of simulations using this nonstandard method for the model (5.8) are presented in Figure 5.4 for different values of the parameter δ. Computer implementation of discrete methods will be discussed in Chapter 8.

5.2 Fixed-Point Analysis

Suppose that $X(t + h) = F(t, X(t))$ is a discrete system with the following difference equations:

$$x_1(t + h) = f_1(t, x_1(t)),$$
$$\vdots \qquad\qquad\qquad\qquad (5.9)$$
$$x_n(t + h) = f_n(t, x_n(t)),$$

where f_i $(i = 1, \ldots, n)$ is a differentiable real-valued function. A point $X^* = (x_1^*, x_2^*, \ldots, x_n^*)$ is called a *fixed point* of the system if $X^* = F(t, X^*)$ for all t. The fixed points of a system of difference equations should correspond to the critical points of the underlying continuous system. In order to analyze the behavior of a discrete system around its fixed points, we apply the fixed-point theorem [48].

Theorem 5.1 (Fixed-point theorem) A fixed point X^* of the discrete system (5.9) is stable if all eigenvalues of the corresponding Jacobian J_{X^*} are less than 1 in magnitude. The fixed point X^* is unstable if at least one of the eigenvalues of J_{X^*} is greater than 1 in magnitude.

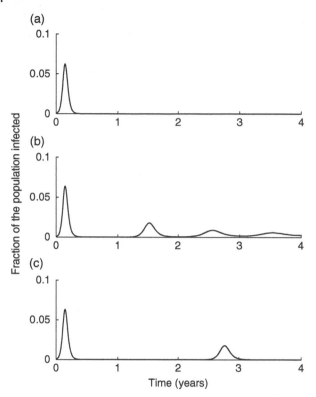

(a)

(b)

(c)

Fraction of the population infected

Time (years)

Figure 5.4 Time profile of the fraction of the population infected. with (a) $\delta = 0$, (b) $\delta = 1/365$, and (c) $\delta = 1/730$. Other parameter values are $\beta = 0.366$ and $\gamma = 0.2439$. Initial population sizes at time $t = 0$ are $I(0) = 0.0005$, $R(0) = 0$, and $S(0) = 1 - I(0)$.

Example 5.5 Let us consider the nonstandard difference equation for the logistic model given by:

$$N(t_{i+1}) = \frac{K(1 + hr)N(t_i)}{K + hrN(t_i)}. \tag{5.10}$$

Clearly, $N = 0$ and $N = K$ are two fixed points of (5.10). The Jacobian of the right-hand side of (5.10) is:

$$J = \frac{(1 + hr)K^2}{(K + hrN)^2}.$$

At $N = 0$, we have $J_0 = 1 + hr$. Since h and r are positive, it follows that $|J_0| > 1$ and therefore $N = 0$ is an unstable fixed point. Thus, if $N(0) > 0$, then the population will never go extinct. At $N = K$, we have $J_K = 1/(1 + hr)$, and hence $|J_K| < 1$. Therefore, $N = K$ is a stable fixed point. As an exercise, investigate how the stability of fixed points in this example relates to the stability of the corresponding critical points of the logistic equation (2.5).

Example 5.6 Consider the predator–prey model (5.3) with the difference equations given by the Euler method in (5.4). It is easy to see that $(x, y) = (0, 0)$ and $(x, y) = (1, 0)$ are two fixed points of (5.4). The Jacobian of (5.4) is:

$$J = \begin{bmatrix} 1 + h[r(1 - 2x) - ay] & -hax \\ hcy & 1 + h(cx - d) \end{bmatrix}.$$

At $(x, y) = (0, 0)$, we have:

$$J_{(0,0)} = \begin{bmatrix} 1 + hr & 0 \\ 0 & 1 - hd \end{bmatrix}.$$

The eigenvalues of $J_{(0,0)}$ are $\lambda_1 = 1 + hr$ and $\lambda_2 = 1 - hd$. Since $|\lambda_1| > 1$, it follows that $(x, y) = (0, 0)$ is an unstable fixed point.

At $(x, y) = (1, 0)$, we have:

$$J_{(1,0)} = \begin{bmatrix} 1 - hr & -ha \\ 0 & 1 + h(c - d) \end{bmatrix},$$

with eigenvalues:

$$\lambda_1 = 1 - hr, \quad \lambda_2 = 1 + h(c - d).$$

If $h < 2/r$, then $|\lambda_1| < 1$, and the stability of $(1, 0)$ depends on the magnitude of λ_2. It is easy to see that $\lambda_2 < 1$ if $c < d$ and $h < -2/(c - d)$. Thus, we have the following conditions for stability of the fixed point $(1, 0)$:

$$c < d \quad \text{and} \quad h < \min\left\{\frac{2}{r}, \frac{-2}{c - d}\right\}.$$

If any of these conditions are violated, then $(1, 0)$ is an unstable fixed point.

The discrete system (5.4) has also a fixed point given by:

$$(x^*, y^*) = \left(\frac{d}{c}, \frac{r(c - d)}{ac}\right),$$

which is located in the region $(x > 0, y > 0)$ if $c > d$ (and this guarantees that $(1, 0)$ is an unstable fixed point). The Jacobian at (x^*, y^*) is:

$$J_{(x^*, y^*)} = \begin{bmatrix} 1 - \dfrac{hrd}{c} & -\dfrac{had}{c} \\ \dfrac{hr(c - d)}{a} & 1 \end{bmatrix}.$$

After some algebraic calculations, we find the characteristic equation of $J_{(x^*, y^*)}$ to be:

$$Q(\lambda) = (1 - \lambda)^2 - A(1 - \lambda) + B = 0,$$

where

$$A = \frac{hrd}{c},$$

$$B = \frac{h^2 rd(c - d)}{c}.$$

Now, we show that if $h > 1/(c - d)$, then (x^*, y^*) is unstable. Suppose $\lambda_{\pm} = u \pm iv$ are complex roots of $Q(\lambda)$, where $v > 0$. We show that $|\lambda_{\pm}| > 1$. Expanding $Q(\lambda_-)$ gives:

$$Q(\lambda_-) = (1 - u + iv)^2 - A(1 - u + iv) + B$$
$$= (1 + u^2 - v^2 - 2u - A + Au + B) - i(2u - 2 + A)v.$$

Since $Q(\lambda_-) = 0$, we have:

$$1 + u^2 - v^2 - 2u - A + Au + B = 0,$$
$$2u - 2 + A = 0.$$

Suppose $u^2 + v^2 \le 1$. Thus, $v^2 < 1 - u^2$, and we have:

$$0 = 1 + u^2 - v^2 - 2u - A + Au + B$$
$$\ge 2u^2 - 2u - A + Au + B$$
$$= -A + B$$
$$= -\frac{hrd}{c} + \frac{h^2rd(c - d)}{c}$$
$$= \frac{hrd}{c}[-1 + h(c - d)] > 0.$$

This contradiction shows that $|\lambda_{\pm}| = u^2 + v^2 > 1$. Suppose now that λ_1 is a real positive root of $Q(\lambda) = 0$. If $0 \le \lambda_1 \le 1$, then:

$$0 = (1 - \lambda_1)^2 - A(1 - \lambda_1) + B > -A + B = \frac{hrd}{c}[-1 + h(c - d)] > 0,$$

which is a contradiction, and hence $\lambda_1 > 1$. Suppose λ_2 is another real root of $Q(\lambda) = 0$. Since $B > 0$, we have $B = (1 - \lambda_1)(1 - \lambda_2) > 0$, which implies that $\lambda_2 > 1$. A similar argument can be used to show that if $\lambda_1, \lambda_2 < 0$ are real numbers, then $\lambda_1 < -1$ and $\lambda_2 < -1$, and therefore $|\lambda_1| > 1$ and $|\lambda_2| > 1$.

Example 5.7 We now consider the same predator–prey model (5.3) with the nonstandard discrete system given by (5.7). It is easy to see that $(x, y) = (0, 0)$, $(x, y) = (1, 0)$, and $(x^*, y^*) = (d/c, r(c - d)/(ac))$ are fixed points of (5.7). The Jacobian of (5.7) is:

$$J = \begin{bmatrix} \dfrac{(1 + hr)(1 + hay)}{(1 + hrx + hay)^2} & \dfrac{-ha(1 + hr)x}{(1 + hrx + hay)^2} \\ \dfrac{hcy}{1 + hd} & \dfrac{(1 + hcx)}{1 + hd} \end{bmatrix}.$$

At $(x, y) = (0, 0)$, we have:

$$J_{(0,0)} = \begin{bmatrix} 1 + hr & 0 \\ 0 & \dfrac{1}{1 + hd} \end{bmatrix}.$$

The eigenvalues of $J_{(0,0)}$ are $\lambda_1 = 1 + hr$ and $\lambda_2 = 1/(1 + hd)$. Since $|\lambda_1| > 1$, it follows that $(x, y) = (0, 0)$ is an unstable fixed point.

At $(x, y) = (1, 0)$, we have:

$$J_{(1,0)} = \begin{bmatrix} \dfrac{1}{1 + hr} & -\dfrac{ha}{1 + hr} \\ 0 & \dfrac{1 + hc}{1 + hd} \end{bmatrix},$$

with eigenvalues:

$$\lambda_1 = \frac{1}{1 + hr}, \quad \lambda_2 = \frac{1 + hc}{1 + hd}.$$

Since $|\lambda_1| < 1$, the stability of $(1, 0)$ depends on the magnitude of λ_2. It is easy to see that $|\lambda_2| < 1$ if and only if $c < d$, in which case $(x, y) = (1, 0)$ is a stable fixed point. If $c > d$, then $(x, y) = (1, 0)$ is an unstable fixed point. From the analysis presented here, we observe that the stability of $(1, 0)$ using the nonstandard method is independent of the time-step h [1]. This is in contrast to the stability analysis using the Euler method in previous example, where the time-step h imposes conditions on the stability of this fixed point.

For the fixed point (x^*, y^*), where $c > d$, we have:

$$J_{(x^*, y^*)} = \begin{bmatrix} 1 - \dfrac{hrd}{c(1 + hr)} & -\dfrac{had}{c(1 + hr)} \\ \dfrac{hr(c - d)}{a(1 + hd)} & 1 \end{bmatrix}.$$

The characteristic equation of $J_{(x^*, y^*)}$ can be written in the form:

$$Q(\lambda) = (1 - \lambda)^2 - A(1 - \lambda) + B = 0,$$

where

$$A = \frac{hrd}{c(1 + hr)},$$

$$B = \frac{h^2 rd(c - d)}{c(1 + hr)(1 + hd)}.$$

Here, we show that if:

$$\Delta = \frac{h(c - d)}{(1 + hd)} - 1 > 0,$$

then (x^*, y^*) is unstable. Suppose $\lambda_{\pm} = u \pm iv$ are complex roots of $Q(\lambda)$, where $v > 0$. We show that $|\lambda_{\pm}| > 1$. Expanding $Q(\lambda_-)$ gives:

$$Q(\lambda_-) = (1 - u + iv)^2 - A(1 - u + iv) + B$$
$$= (1 + u^2 - v^2 - 2u - A + Au + B) - i(2u - 2 + A)v.$$

Since $Q(\lambda_-) = 0$, we have:

$$1 + u^2 - v^2 - 2u - A + Au + B = 0,$$
$$2u - 2 + A = 0.$$

Suppose $u^2 + v^2 \leq 1$. Thus, $v^2 < 1 - u^2$, and we have:

$$0 = 1 + u^2 - v^2 - 2u - A + Au + B$$
$$\geq 2u^2 - 2u - A + Au + B$$
$$= -A + B$$
$$= -\frac{hrd}{c(1+hr)} + \frac{h^2rd(c-d)}{c(1+hr)(1+hd)}$$
$$= \frac{hrd}{c(1+hr)}\left(-1 + \frac{h(c-d)}{(1+hd)}\right)$$
$$= \left(\frac{hrd}{c(1+hr)}\right)\Delta > 0.$$

This contradiction shows that $|\lambda_\pm| = u^2 + v^2 > 1$. Suppose now that λ_1 is a real positive root of $Q(\lambda) = 0$. If $0 \leq \lambda \leq 1$, then:

$$0 = (1 - \lambda_1)^2 - A(1 - \lambda_1) + B > -A + B = \left(\frac{hrd}{c(1+hr)}\right)\Delta > 0,$$

which is a contradiction, and hence $\lambda_1 > 1$. Suppose λ_2 is another real root of $Q(\lambda) = 0$. Since $B > 0$, we have $(1 - \lambda_1)(1 - \lambda_2) > 0$, which implies that $\lambda_2 > 1$. A similar argument can be used to show that if $\lambda_1, \lambda_2 < 0$, then $\lambda_1 < -1$ and $\lambda_2 < -1$, and therefore $|\lambda_1| > 1$ and $|\lambda_2| > 1$.

Exercises

1   Consider the first-order differential equation $y' = y^2 - x$ with the initial condition $y(0) = -1$. Estimate $y(1)$ and $y(2)$ using the Euler method with time-step $h = 0.2$.

2   Consider the following system of equations for the spread of a disease in a population:

$$S' = \Lambda - \frac{\beta SI}{N} - \mu S,$$
$$I' = \frac{\beta SI}{N} - (\gamma + \mu)I,$$
$$R' = \gamma I - \mu R,$$

where S, I, and R are the population sizes of susceptible, infected, and recovered individuals with $N = S + I + R$, Λ is the constant recruitment rate, β is the transmission rate, γ is the recovery rate of infected individuals, and μ is the natural death rate. For this model, develop both Euler and nonstandard methods, and discuss the stability of the fixed points in each method.

3   Consider the following predator–prey model [33]:

$$x' = rx\left(1 - \frac{x}{K}\right) - \phi(x)y,$$
$$y' = (\mu\phi(x) - d)y.$$

Develop Euler and nonstandard methods to obtain systems of difference equations when:

a) $\phi(x) = \frac{mx}{1+x}$,

b) $\phi(x) = \frac{mx^2}{1+x^2}$,

where all parameters r, K, μ, d, and m are positive. Discuss the stability of the fixed points for each method.

4   Suppose a difference equation has the form:

$$x(t + 1) = 1 + \frac{1}{1 + x(t)}, \quad x(t) \neq -1.$$

Find the fixed points of this equation and discuss their stability.

5   Write a set of difference equations corresponding to the following system of differential equations and analyze the fixed points:

$$x' = 2x - y - x^2,$$
$$y' = x - 2y + y^2.$$

6

Probability and Random Variables

In Chapter 1 we summarized the classification of models according to their nature as deterministic or stochastic. In previous chapters, we considered and analyzed various examples of deterministic models. In order to introduce stochastic modelling, we need to understand the concept of randomness. In phenomena that involve randomness, the system outcomes cannot be predicted with certainty. This is mainly due to the fact that some events underlying the system dynamics occur randomly. From previous examples, we see that the randomness and uncertainty are absent in deterministic models. In this chapter we briefly review the concepts of probability and random variables [43], and develop the probabilistic characterization of random variables as the basis for stochastic modelling.

6.1 Basic Concepts

We begin by introducing some of the basic concepts of probability theory and defining the associated terminology relating to random experiments whose outcomes are not predictable.

Definition 6.1 (Sample Space)   The set of all possible outcomes of an experiment is called the *sample space*.

For example, in flipping a coin, there are two possible outcomes, heads (H) and tails (T), which gives the sample space:

$$S = \{H, T\}.$$

If the experiment is rolling two dice simultaneously, then the sample space has 36 elements and is given by:

$$S = \{(1, 1), (1, 2), \dots, (1, 6), \dots, (6, 1), (6, 2), \dots, (6, 6)\}.$$

Mathematical Modelling: A Graduate Textbook, First Edition. Seyed M. Moghadas and Majid Jaberi-Douraki.
© 2019 John Wiley & Sons, Inc. Published 2019 by John Wiley & Sons, Inc.
Companion Website: www.wiley.com/go/Moghadas/Mathematicalmodelling

For the experiment of flipping two coins, we have the sample space:

$$S = \{(H,H),(H,T),(T,H),(T,T)\}.$$

Definition 6.2 (Event) Any subset $E \subseteq S$ is called an *event*.

In the example of rolling two dice, the event in which the sum of numbers facing up is 7 includes:

$$E = \{(1,6),(2,5),(3,4),(4,3),(5,2),(6,1)\}.$$

☞ **Properties of events.** Suppose E and F are two events. We have the following properties:

1) $E \cup F$ is also an event and occurs when E or F occurs.
2) EF is also an event and occurs when both E and F occur.
3) $\emptyset$ is referred to a *null event* with no outcomes. Simply, $\emptyset$ cannot occur.
4) If $EF = \emptyset$, then E and F are said to be *mutually exclusive* events.
5) E^c is called the *complement* of E and refers to an event that has all the outcomes in sample space except those that are contained in E. Thus, $E \cup E^c = S$ and $EE^c = \emptyset$.

☞ **Probability.** Occurrences of events are associated with numerical values. Suppose a random experiment is repeated n times. If the event E occurs n_E times, then the probability of E, denoted by $P(E)$, is defined by:

$$P(E) = \lim_{n \to \infty} \frac{n_E}{n}.$$

This probability is defined as a relative frequency. In the classical definition, when the total number of experiments N is finite, the probability of E is defined by:

$$P(E) = \frac{N_E}{N},$$

where N_E is the number of outcomes favorable to the event E. The probability of any event in the sample space obeys the following axioms:

1) $0 \leq P(E) \leq 1$.
2) $P(S) = 1$, and therefore $P(\emptyset) = 0$.
3) For any sequence $E_1, E_2, \ldots$ of mutually exclusive events, that is, $E_m E_n = \emptyset$ if $m \neq n$, we have:

$$P\left(\bigcup_{i=1}^{\infty} E_i\right) = \sum_{i=1}^{\infty} P(E_i).$$

We should note that for any two events E and F, we have:

$$P(E \cup F) = P(E) + P(F) - P(EF).$$

In the example of flipping a fair coin, the probabilities of events $E = \{H\}$ and $F = \{T\}$ are the same and $P(\{H\}) = P(\{T\}) = \frac{1}{2}$. In rolling two dice, there are six outcomes for which the sum of two numbers is 7. Therefore $N_E = 6$. The total number of outcomes is $N = 36$, which is the same as the number of elements in the sample space. Thus,

$$P(E) = P(\{(1,6),(2,5),(3,4),(4,3),(5,2),(6,1)\}) = \frac{6}{36} = \frac{1}{6}.$$

Suppose we flip two fair coins simultaneously. The sample space has four elements, corresponding to the number of possible outcomes. Let:

$$E = \{(H,T),(H,H)\}, \quad F = \{(H,T),(T,T)\}.$$

Thus, $EF = \{(H,T)\}$ and the probability of $E \cup F$ is:

$$P(E \cup F) = \frac{2}{4} + \frac{2}{4} - \frac{1}{4} = \frac{3}{4}.$$

6.2 Conditional Probabilities

In random experiments, it is possible that the occurrence of an event depends on (i.e., is conditional on) the occurrence of another event. Suppose E and F are two events and $P(F) > 0$. The probability that E occurs given that F has occurred is denoted by $P(E|F)$ and defined as the conditional probability of E. If F occurs, then for E to occur, it is necessary for the actual occurrence (or the outcome) to be a point in both E and F. Therefore, the outcome must be in EF. Since F has occurred, F is the new sample space for E. This means that E is part of EF (Figure 6.1). Hence, the probability of E occurring in this new sample space is:

$$P(E|F) = \frac{P(EF)}{P(F)}.$$

Example 6.1 Suppose we have a hat containing 12 balls numbered from 1 to 12. We draw a ball from the hat. If this draw is a number greater than 5, we are

Figure 6.1 Representation of sample spaces for the conditional probability of $P(E|F)$.

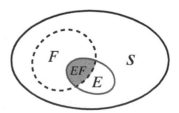

interested in finding the probability that it is the ball with number 8. This is a conditional probability. In this case, F is the event that the ball has a number greater than 5. Thus, $P(F) = \frac{7}{12}$. If E is the event that the ball has number 8 in this draw, then $P(EF) = \frac{1}{12}$. Therefore,

$$P(E|F) = \frac{\frac{1}{12}}{\frac{7}{12}} = \frac{1}{7}.$$

☞ **Independent events.** Two events are said to be independent if:

$$P(EF) = P(E)P(F).$$

In this case, we have:

$$P(E|F) = \frac{P(E)P(F)}{P(F)} = P(E).$$

Example 6.2   Consider the rolling of two dice with the following events:

i) E, the sum of the two numbers is 6;
ii) F, the first die is 4.

These events are:

$$E = \{(1, 5), (2, 4), (3, 3), (4, 2), (5, 1)\},$$
$$F = \{(4, 1), (4, 2), (4, 3), (4, 4), (4, 5), (4, 6)\}.$$

Thus, $EF = \{(4, 2)\}$ and,

$$P(EF) = P(\{(4, 2)\}) = \frac{1}{36}.$$

We also have:

$$P(E)P(F) = \frac{5}{36} \times \frac{1}{6} = \frac{5}{216}.$$

Since $P(EF) \neq P(E)P(F)$, the two events are not independent. If we change the event E so that it is now "the sum of the two numbers is 7", then:

$$E = \{(1, 6), (2, 5), (3, 4), (4, 3), (5, 2), (6, 1)\},$$

and $EF = \{(4, 3)\}$. Thus,

$$P(EF) = P\{(4, 3)\} = \frac{1}{36},$$

and

$$P(E) = P\{(1, 6), (2, 5), (3, 4), (4, 3), (5, 2), (6, 1)\} = \frac{6}{36} = \frac{1}{6}.$$

This implies that $P(EF) = P(E)P(F) = \frac{1}{36}$, and therefore events E and F are independent.

6.3 Random Variables

A random variable X is a real-valued function defined on the sample space S, that is, $X : S \to \mathbb{R}$ (Figure 6.2).

Consider the example of rolling two dice, with the outcome that the sum of two numbers is 7. The associated event is:

$$E = \{(1, 6), (2, 5), (3, 4), (4, 3), (5, 2), (6, 1)\}.$$

We can define a random variable that takes E to number 7, that is $X(E) = 7$. Thus,

$$P(X = 7) = \frac{6}{36} = \frac{1}{6}.$$

If a random variable takes a finite or countable number of possible values in $\mathbb{R}$, then it is called *discrete*. If a random variable takes continuous values of some subintervals in $\mathbb{R}$, then it is called *continuous*.

6.3.1 Cumulative Distribution Function

A cumulative distribution function (CDF) is a real-valued function $F(\cdot)$ of a random variable X defined by:

$$F(b) = P(X \le b), \quad b \in (-\infty, \infty).$$

Thus, $F(b)$ denotes the probability that the random variable X takes on a value less than or equal to b.

☞ **Properties of CDFs.** A CDF $F(\cdot)$ has the following properties:

1) $F(b) \ge 0$ for all b, and F is a nondecreasing function of b. This is evident from the fact that if $a < b$, then $\{X \le a\}$ is contained in $\{X \le b\}$.
2) $\lim_{b \to +\infty} F(b) = 1$, and $\lim_{b \to -\infty} F(b) = 0$.
3) For a random variable $a < X \le b$,

$$P(X \le b) - P(X \le a) = F(b) - F(a).$$

Figure 6.2 Representation of a random variable defined on the sample space S.

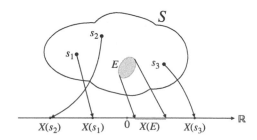

We should note that:

$$P(X < b) = \lim_{h \to 0^+} P(X \le b - h) = \lim_{h \to 0^+} F(b - h).$$

However, it is not necessarily the case that $\lim_{h \to 0^+} F(b - h) = F(b)$.

Example 6.3 Suppose X is defined as the sum of two numbers when rolling two dice. We are interested in finding $P(3 < X \le 5)$. The associated event is:

$$E = \{(1, 3), (1, 4), (2, 2), (2, 3), (3, 1), (3, 2), (4, 1)\},$$

and therefore,

$$P(3 < X \le 5) = \frac{7}{36}.$$

We note that:

$$F(5) = P(X \le 5) = P(\{(1, 1), (1, 2), (1, 3), (1, 4), (2, 1),$$

$$(2, 2), (2, 3), (3, 1), (3, 2), (4, 1)\}) = \frac{5}{18},$$

and

$$F(3) = P(X \le 3) = P(\{(1, 1), (1, 2), (2, 1)\}) = \frac{3}{36} = \frac{1}{12}.$$

Thus, $F(5) - F(3) = \frac{7}{36}$.

6.3.2 Discrete Random Variables

For a discrete random variable X, we define a real-valued function $p : \mathbb{R} \to [0, 1]$ by $p(a) = P(X = a)$. This function is called the *probability mass function* (PMF). If X takes the values $x_1, x_2, \ldots$, then:

$$p(x) = \begin{cases} > 0 & \text{if } x = x_i \\ 0 & \text{if } x \ne x_i \end{cases}.$$

Since X covers the entire sample space, $\sum_i p(x_i) = 1$. For a discrete random variable, the CDF can be expressed in terms of the PMF as:

$$F(a) = \sum_{x_i \le a} p(x_i).$$

Example 6.4 Suppose X has a PMF given by $p(1) = \frac{1}{2}, p(2) = \frac{1}{3}$, and $p(3) = \frac{1}{6}$. Thus, the CDF is given by:

$$F(a) = \begin{cases} 0, & a < 1, \\ \frac{1}{2}, & 1 \le a < 2, \\ \frac{5}{6}, & 2 \le a < 3, \\ 1, & 3 \le a. \end{cases}$$

☞ **Specific discrete random variables.** There are a number of discrete random variables that are widely used in the study of systems involving random experiments. Here, we provide some of these random variables.

Bernoulli. If X is a random variable with $P(X = 1) = p$ and $P(X = 0) = 1 - p$, for some $0 < p < 1$, then X is called a Bernoulli random variable. This random variable is often used for experiments with outcomes that are deemed as either success or failure. In this way, $X = 1$ is regarded as success and $X = 0$ reflects failure. With this random variable, the PMF is defined by $p(1) = p$ and $p(0) = 1 - p$.

Binomial. When n experiments are conducted with success or failure outcomes, and the probability p of success is the same in each experiment, we have a binomial random variable X with parameters (n, p). In this case, the PMF is given by:

$$p(i) = \binom{n}{i} p^i (1 - p)^{n-i}, \quad i = 0, 1, 2, \ldots, n,$$

with

$$\binom{n}{i} = \frac{n!}{i!(n - i)!},$$

where i is the number of successes in n trials.

Geometric. Often, success is a determining factor for halting an experiment. When experiments are conducted until one success occurs, we have a geometric random variable. If success occurs on the nth trial, then $X = 1, 2, \ldots, n - 1$ correspond to failure, and $X = n$ reflects the success. In this case, the PMF is $p(n) = p(1 - p)^{n-1}$.

Poisson. Sometimes the desired outcome of an experiment has a small probability of occurrence and therefore a large number of independent trials may be needed to achieve this outcome. Suppose λ is the average number of occurrences over a specific period of time. Then, the Poisson random variable X counts the total number of occurrences during a given time period. Thus, X takes one of the values $0, 1, 2, 3, \ldots$ and its PMF is defined by:

$$p(i) = P(X = i) = e^{-\lambda} \frac{\lambda^i}{i!},$$

where i is the number of occurrences.

Example 6.5 In this example, we are interested in determining the probability of hitting different parts of a dartboard as shown in Figure 6.3 based on the areas between the circles or the main target, which is the smallest circle in the middle. The largest circle representing the dartboard has radius r, and the smaller inner circles have radii of:

$$r_1 = \frac{3}{4}r, \quad r_2 = \frac{1}{2}r, \quad r_3 = \frac{1}{4}r. \tag{6.1}$$

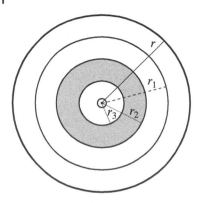

Figure 6.3 Representation of a dartboard.

The target circle has radius $r_4 = r/12$. Suppose, when throwing a dart, the probability of hitting anywhere on the dartboard is 0.8, which means that there is a 20% chance of the dart landing outside the dartboard. Also, suppose that the probability of hitting the dartboard between two circles or the target circle is the same as the proportional area. For example, the probability of hitting the gray area illustrated in Figure 6.3 is the ratio of the area between the circles with radii r_2 and r_3 to the entire area of the dartboard. This probability is given by:

$$P = 0.8 \left(\frac{\pi(r_2^2 - r_3^2)}{\pi r^2} \right) = \frac{2.4}{16} = 0.15.$$

The probability of hitting the target circle with radius r_4 is $P_c = \frac{0.8}{144}$. We now determine the probability of hitting the target circle exactly once in n throws. This corresponds to a binomial random variable, where the probability of one hit in n throws is:

$$P_n(1) = \binom{n}{1} P_c (1 - P_c)^{n-1} = \frac{0.8n}{144} \left(\frac{143.2}{144} \right)^{n-1}.$$

We note that $\lim_{n \to \infty} P_n(1) = 0$. Considering P_n as a function of n and taking the derivative, we find that:

$$P_n'(1) = \frac{0.8}{144} \left(\frac{143.2}{144} \right)^{n-1} + \frac{0.8n}{144} \ln \left(\frac{143.2}{144} \right) \left(\frac{143.2}{144} \right)^{n-1},$$

which has the root $-1/\ln(143.2/144) \approx 179.4995$. Thus, the highest probability of hitting the target circle exactly once corresponds to 179 throws, which is $P_{179}(1) = 0.3689$. Figure 6.4 shows the probability $P_n(1)$ for different values of n. As an exercise, determine the probability $P_n(2)$ for the gray area in Figure 6.3.

6.3.3 Continuous Random Variables

Suppose, for a continuous random variable X, that there exists a nonnegative function $f(x)$, defined for all $x \in (-\infty, +\infty)$, with the property that for any set $B \subseteq \mathbb{R}$,

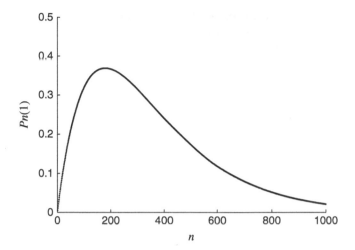

Figure 6.4 Probability of hitting the gray area on the dartboard in Figure 6.3 exactly once as a function of the number of throws.

Figure 6.5 Representation of a probability density function.

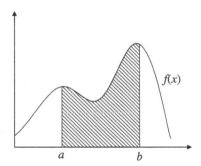

$$P(X \in B) = \int_B f(x)\mathrm{d}x.$$

Then f is called the *probability density function* (PDF). Unlike the case of discrete random variables, the PDF is used to specify the probability of a continuous random variable falling within a particular range of values, as opposed to taking any one value (Figure 6.5).

For any PDF,

$$P(X \in (-\infty, +\infty)) = \int_{-\infty}^{\infty} f(x)\mathrm{d}x = 1.$$

Similar to the case of discrete random variables, there is a relation between the PDF and CDF $F(\cdot)$, given by:

$$F(a) = P(X \in (-\infty, a)) = \int_{-\infty}^{a} f(x)\mathrm{d}x.$$

☞ **Specific continuous random variables.** There are several continuous random variables that are commonly used in stochastic processes. Here we provide the functional form of the PDF for some of these random variables.

Uniform. The PDF for a uniform random variable distributed over $[0, 1]$ is:

$$f(x) = \begin{cases} 1, & 0 \leq x \leq 1, \\ 0, & \text{otherwise.} \end{cases}$$

If the random variable takes on values in $(\alpha, \beta) \subseteq \mathbb{R}$, then the PDF is:

$$f(x) = \begin{cases} \frac{1}{\beta - \alpha}, & \alpha \leq x \leq \beta, \\ 0, & \text{otherwise.} \end{cases}$$

Exponential. For a positive $\lambda > 0$, the exponential PDF is:

$$f(x) = \begin{cases} \lambda e^{-\lambda x}, & \text{if } x \geq 0, \\ 0, & \text{if } x < 0. \end{cases}$$

The parameter $1/\lambda$ is the expected value (or mean) of the exponential random variable X, which is calculated as follows:

$$E(X) = \int_{-\infty}^{+\infty} x f(x) \mathrm{d}x = \int_{0}^{+\infty} \lambda x e^{-\lambda x} \mathrm{d}x = \frac{1}{\lambda}.$$

Gamma. For positive parameters λ, $\alpha > 0$, the gamma PDF has the form:

$$f(x) = \begin{cases} \frac{\lambda e^{-\lambda x}(\lambda x)^{\alpha-1}}{\Gamma(\alpha)}, & \text{if } x \geq 0, \\ 0, & \text{if } x < 0, \end{cases}$$

where the Γ function is defined by:

$$\Gamma(\alpha) = \int_{0}^{+\infty} e^{-x} x^{\alpha-1} \mathrm{d}x.$$

The expected value of the gamma random variable X, is:

$$E(X) = \int_{-\infty}^{+\infty} x f(x) \mathrm{d}x = \int_{0}^{+\infty} \frac{e^{-\lambda x}(\lambda x)^{\alpha}}{\Gamma(\alpha)} \mathrm{d}x = \frac{\alpha}{\lambda}.$$

Normal. For a random variable X that is normally distributed with expected value μ and variance σ^2, the PDF is:

$$f(x) = \frac{1}{\sqrt{2\pi\sigma^2}} e^{\frac{-(x-\mu)^2}{2\sigma^2}},$$

where $x \in \mathbb{R}$. The variance, σ^2, measures the extent to which values taken by the random variable are spread out from their expected value μ.

6.3.4 Waiting Time

Events of the same nature that occur individually at random moments may occur at an average rate when considered as a group, which corresponds to a Poisson process. The waiting time between the occurrence of events in a Poisson process can generally be described by an exponential distribution. Suppose that T is the amount of time elapsed after a Poisson event, and let $P(T > t)$ be the probability that no event has occurred in the time internal $[0, t]$. If the events have λ as the average rate of occurrence, then $P(T > t) = e^{-\lambda t}$. This is obtained by letting $i = 0$ (i.e., no events occur in $[0, t]$) in the PMF of the Poisson process. Thus, the CDF (i.e., the probability that at least one event has occurred in the time interval $[0, t]$) is:

$$F(t) = P(T \le t) = 1 - e^{-\lambda t}.$$

Differentiating gives us the PDF for the waiting time between the occurrence of events:

$$f(t) = F'(t) = \lambda e^{-\lambda t}, \qquad t \ge 0.$$

Example 6.6 After graduation, a group of students decide to celebrate by having a gathering in a restaurant. All the students arrive at the restaurant within 1 hour at the rate of 5 students every 15 minutes. We are interested in determining the probability that no student arrives in 5 minutes. If we assume that the arrival is a Poisson process with the mean of $\lambda = 5/15$ students per minute, then the waiting time has the PDF:

$$f(t) = \frac{1}{3}e^{-t/3}, \qquad t \ge 0.$$

Thus, the probability of waiting more than 5 minutes for the arrival of a student is:

$$P(\text{arrival} > 5 \text{ minutes}) = 1 - P(\text{arrival} \le 5 \text{ minutes})$$

$$= 1 - \int_0^5 f(t)dt$$

$$= 1 - \frac{1}{3}\int_0^5 e^{-t/3}dt = e^{-\frac{5}{3}} \approx 0.189.$$

Exercises

1 Suppose two dice are rolled consecutively. If the first die results in a 4, what is the chance that the sum of two numbers is 9 once the second die is rolled?

2 What is the probability that the sum of numbers in rolling two dice will be greater than 8, given that the first die is a 5?

3 Suppose E and F are two events with $P(E) > 0$ and $P(F) > 0$. Show that if $P(E|F) = P(E)$, then $P(F|E) = P(F)$.

4 For the probability mass functions in discrete random variables, show that:
a) If the random variable is geometric, then $\sum_{n=1}^{\infty} p(n) = 1$.
b) If the random variable is Poisson, then $\sum_{i=0}^{\infty} p(i) = 1$.

5 Suppose births in a hospital occur randomly, with an associated random variable that has a Poisson distribution. Suppose births occur at an average rate of 2 per hour.
a) What is the probability of observing exactly 4 births in a given hour?
b) What is the probability of observing 2 or more births in a given hour?

6 Suppose simulation jobs arrive at the computing system every 20 seconds on average. What is the probability that the waiting time to receive a job is no more than 40 seconds?

7 Hospital records show that the risk of death for patients with a specific comorbid illness is 15% after hospitalization. What is the probability that of six randomly selected patients, four will recover?

8 An auto dealer sells 10 cars on average per week. Find the probability that in a given week the dealer will sell at least 8 but no more than 14 cars.

9 Using the properties of the gamma function Γ, show that, for $\alpha, \lambda > 0$,

$$\int_0^{\infty} \frac{(\lambda x)^{\alpha-1} e^{-\lambda x}}{\Gamma(\alpha)} dx = \frac{1}{\lambda}.$$

7

Stochastic Modelling

The word *stochastic* originates from the Greek word $\sigma\tau o\chi\alpha\sigma\tau\iota\kappa\acute{o}\varsigma$ (meaning "random" or "chance"). When a phenomenon is free of any randomness, the outcomes are certain and we can predict them by using deterministic models. However, when the occurrence of outcomes is associated with probabilities, then we need a stochastic model to predict a set of possible outcomes. In this chapter, we describe some key components of stochastic models [55], with examples of their applications.

In a stochastic model, the probability of the system being in different states may change over time. To better understand the conceptual framework of stochastic models, let us consider a simple example of population growth. Suppose $r > 0$ is the average rate of exponential growth in the population. Then, a simple discrete equation for population size n over time is given by:

$$n(t + 1) = rn(t).$$

Using this equation, if $n(t)$ is known, then we can precisely predict the population size at time $t + 1$. Realistically, however, the rate of growth may be different at different times. In a stochastic model, this growth rate can be sampled (i.e., randomly selected) from an exponential distribution with mean r. Thus, we may observe different population sizes at time $t + 1$, depending on the rate r. Therefore, in a stochastic model, there is a probability associated with observing a specific population size at time $t + 1$.

7.1   Stochastic Processes

Definition 7.1 A stochastic process is a collection of random variables $\{X_t(s) : t \in T, s \in S\}$, where t is in an index set T (usually representing time) and S is the sample space of random variables. For each fixed t, $X_t(s)$ corresponds to a function defined on T that is called a *stochastic realization* of the process. For simplicity of notation, we may denote the stochastic realization by $X_t = X(t)$, omitting s.

Mathematical Modelling: A Graduate Textbook, First Edition. Seyed M. Moghadas and Majid Jaberi-Douraki.
© 2019 John Wiley & Sons, Inc. Published 2019 by John Wiley & Sons, Inc.
Companion Website: www.wiley.com/go/Moghadas/Mathematicalmodelling

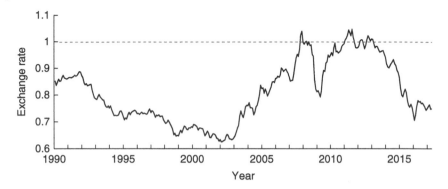

Figure 7.1 Monthly exchange rate of the Canadian dollar against the US dollar over the period 1990–2017.

An example of a stochastic process is the exchange rate of the Canadian dollar against the US dollar; Figure 7.1 shows the exchange rate on monthly basis over a 27-year period. In this case, the random variable is the ratio of the Canadian dollar to the US dollar, which changes with time.

We now proceed with the classic example of the birth and death model to provide a better understanding of stochastic processes. Let us first study the birth model.

Example 7.1 (Birth model) Suppose an organism reproduces at a constant rate b per unit time. Since b is constant, it takes time $1/b$ (on average) for the organism to reproduce. With this exponentially distributed waiting time, we have:

$$n(t + \Delta t) - n(t) = bn(t)\Delta t.$$

In a deterministic context, when $\Delta t \to 0$, we obtain $n(t) = n_0 e^{bt}$, where $n_0 = n(0)$, and therefore the population size is known at any time t. In a stochastic context, the population size is associated with a probability. Let $X_t = X(t)$ be the random variable denoting the population size at time t. For any $n = 0, 1, 2, \ldots$, let $p_n(t)$ be the probability mass function at time t, given by:

$$p_n(t) = P(X_t = n).$$

We assume that X_t has the following properties for a small amount of time Δt:

1) The probability of more than one birth in Δt units of time is negligible. Mathematically, we express this as:

$$\lim_{\Delta t \to 0} \frac{\mathcal{O}(\Delta t^2)}{\Delta t} = 0,$$

and therefore, for a sufficiently small time interval Δt, the probability that a birth occurs in a population of size n is:

$nb\Delta t + \mathcal{O}(\Delta t^2).$

2) At $t = 0$, $P(X_0 = a) = 1$. This represents the initial size of the population.

Suppose $n \geq 1$. Then the probability $p_n(t + \Delta t) = P(X_{t+\Delta t} = n)$ has the following components:

1) At time t, $X_t = n - 1$ and a birth has occurred during $(t, t + \Delta t]$. Thus,

$$p_n^{(1)}(t + \Delta t) = p_{n-1}(t)((n-1)b\Delta t + \mathcal{O}(\Delta t^2)), \tag{7.1}$$

where $p_{n-1}(t) = P(X_t = n - 1)$.
2) At time t, $X_t = n$ and no birth has occurred during $(t, t + \Delta t]$. Thus,

$$p_n^{(2)}(t + \Delta t) = p_n(t)(1 - nb\Delta t - \mathcal{O}(\Delta t^2)), \tag{7.2}$$

where $p_n(t) = P(X_t = n)$, and $1 - nb\Delta t - \mathcal{O}(\Delta t^2)$ is the probability of no birth occurring.

With these components, $P(X_{t+\Delta t} = n)$ is the same as the probability that one of the following outcomes occurs:

1) E_1 (birth): $X_t = n - 1$ and $X_{t+\Delta t} = n$.
2) E_2 (no birth): $X_t = n$ and $X_{t+\Delta t} = n$.

This implies that:

$$P(X_{t+\Delta t} = n) = P(E_1 \cup E_2) = P(E_1) + P(E_2) - P(E_1 E_2).$$

Since E_1 and E_2 are mutually exclusive events, we have $P(E_1 E_2) = 0$. Thus,

$$P(X_{t+\Delta t} = n) = P(E_1) + P(E_2)$$
$$= P(X_{t+\Delta t} = n | X_t = n - 1) + P(X_{t+\Delta t} = n | X_t = n).$$

Adding (7.1) and (7.2) gives:

$$p_n(t + \Delta t) = p_{n-1}(t)(n-1)b\Delta t + p_n(t)(1 - nb\Delta t) + \mathcal{O}(\Delta t^2).$$

Rearranging this equation and dividing by Δt, we get:

$$\frac{p_n(t + \Delta t) - p_n(t)}{\Delta t} = p_{n-1}(t)b(n-1) - p_n(t)bn + \frac{\mathcal{O}(\Delta t^2)}{\Delta t}. \tag{7.3}$$

Taking the limit of (7.3) as $\Delta t \to 0$ gives:

$$\frac{dp_n(t)}{dt} = p_{n-1}(t)(n-1)b - p_n(t)nb. \tag{7.4}$$

This is referred to as the "master equation" for the birth model, which provides the change in the PMF associated with the population size $X_t = n$. This equation is valid for $n \geq 1$.

When $n = 0$, the event E_1 cannot occur, and therefore E_2 occurs (no birth). This means that $p'_0(t) = 0$ for all t, and therefore p_0 is a fixed probability. This fixed probability can be determined at equilibrium using the critical points of (7.4). This is obtained by solving:

$$p_{n-1}(n-1)b - p_n nb = 0, \qquad (7.5)$$

If $n = 1$, then $p_1 b = 0$, and since $b > 0$, we have $p_1 = 0$. If $n = 2$, then we have $p_1 b - 2p_2 b = 0$. It follows from $p_1 = 0$ that $p_2 = 0$. An induction argument shows that $p_n = 0$ for $n \geq 1$. Since $\sum_{n \geq 0} p_n = 1$, we find that $p_0 = 1$, and therefore the population remains at $n = 0$. We can extract further information regarding the population size from (7.4). Suppose $n = 1$ at time $t = 0$, and therefore $p_1(0) = 1$. From (7.4), we have:

$$\frac{dp_1(t)}{dt} = -p_1(t)b.$$

Solving this equation gives $p_1(t) = e^{-bt}$, which implies that $p_1(t) \to 0$ as $t \to \infty$. This suggests that the probability of the population size remaining at $n = 1$ decreases with t, and therefore the population will grow in size greater than 1.

Example 7.2 (Birth and death model)   Suppose an organism produces at a constant rate b and dies at a constant rate d. The waiting times for these events are $1/b$ and $1/d$ (on average). In a deterministic form, the change in population size can be formulated as:

$$n(t + \Delta t) - n(t) = bn(t)\Delta t - dn(t)\Delta t = (b-d)n\Delta t.$$

Solving the resulting differential equation when $\Delta t \to 0$ gives $n(t) = n_0 e^{(b-d)t}$, where n_0 is the initial population size at time $t = 0$. If $b > d$, then the population will grow unboundedly. If $b < d$, then the population will go extinct over time. If $b = d$, then the population remains at n_0 for all t.

In a stochastic model, since birth and death occur randomly, the population size is associated with a probability. Similar to the birth model, let $X_t = X(t)$ be the random variable denoting the population size at time t. For sufficiently small Δt, the probability of a death occurring in a population of size n is $nd\Delta t + \mathcal{O}(\Delta t^2)$. In this model, the probability mass function $p_n(t)$ has the following three components for $n \geq 1$:

1) At time t, $X_t = n - 1$ and only a birth has occurred during $(t + \Delta t]$. Thus,

$$p_n^{(1)}(t + \Delta t) = p_{n-1}(t)(n-1)b\Delta t + \mathcal{O}(\Delta t^2).$$

2) At time t, $X_t = n + 1$ and only a death has occurred during $(t + \Delta t]$. Thus,

$$p_n^{(2)}(t + \Delta t) = p_{n+1}(t)(n+1)d\Delta t + \mathcal{O}(\Delta t^2).$$

3) At time t, $X_t = n$ and no birth or death has occurred during $(t + \Delta t]$. Thus,

$$p_n^{(3)}(t + \Delta t) = p_n(t)(1 - nb\Delta t - \mathcal{O}(\Delta t^2))(1 - nd\Delta t - \mathcal{O}(\Delta t^2)).$$

We also note that the occurrence of more than one event (i.e., both birth and death) during $(t, t + \Delta t]$ has probability $\mathcal{O}(\Delta t^2)$. With the above components, $P(X_{t+\Delta t} = n)$ is the same as the probability that one of the following events occurs:

1) E_1 (birth): $X_t = n - 1$ and $X_{t+\Delta t} = n$.
2) E_2 (death): $X_t = n + 1$ and $X_{t+\Delta t} = n$.
3) E_3 (no birth and no death): $X_t = n$ and $X_{t+\Delta t} = n$.

Here, we have implicitly made the assumption that during the Δt period of time, only one event can occur (which holds true if Δt is sufficiently small), and therefore disregard the possibility of having both a birth and a death. Considering the above events, we have:

$$P(X_{t+\Delta t} = n) = P(E_1 \cup E_2 \cup E_3) = P(E_1) + P(E_2) + P(E_3),$$

where we have used the property of mutually exclusive events. Adding these probabilities gives:

$$p_n(t + \Delta t) = p_{n-1}(t)(n - 1)b\Delta t$$
$$+ p_n(t)(1 - nb\Delta t - nd\Delta t)$$
$$+ p_{n+1}(t)(n + 1)d\Delta t + \mathcal{O}(\Delta t^2).$$

Rearranging this equation and taking the limit as $\Delta t \to 0$, we get:

$$\frac{dp_n(t)}{dt} = p_{n-1}(t)(n - 1)b - p_n(t)n(b + d) + p_{n+1}(t)(n + 1)d, \qquad (7.6)$$

which represents the master equation for the birth and death model.

When $n = 0$, no birth or death can occur and therefore $p_0'(t) = p_1(t)d$. The critical points of the model are obtained by solving:

$$0 = p_{n-1}(n - 1)b - p_n n(b + d) + p_{n+1}(n + 1)d, \quad n \geq 1,$$
$$0 = p_1 d, \quad n = 0.$$

Thus, $p_1 = 0$, and hence $2p_2 d = 0$ (for the case $n = 1$), which implies that $p_2 = 0$. Repeating this process, we find that $p_n = 0$ for $n \geq 1$. Since $\sum_{n \geq 0} p_n = 1$, we find that $p_0 = 1$. Therefore, $p_0 = 1$ and $p_n = 0$ for $n \geq 1$ is a critical point of the model.

Now suppose that $0 \leq p_0 < 1$. Because $p_1 = 0$, we find $p_n = 0$ for $n \geq 1$. Therefore $\sum_{n \geq 0} p_n = p_0 < 1$, which is impossible according to the properties of the PMF. This shows that if $0 \leq p_0 < 1$, then the model has no critical points.

7.2 Probability Generating Function

Definition 7.2 For a discrete random variable X taking values in the nonnegative integers $0, 1, 2, \ldots$, the probability generating function is defined by:

$$G(z) = \sum_{i=0}^{\infty} p_i z^i, \tag{7.7}$$

where $p_i = P(X = i)$ is the PMF and z is a complex number.

When $\|z\| < 1$, since $0 \le p_i \le 1$, we have:

$$\|G(z)\| \le \sum_{i=0}^{\infty} \|z\|^i = \frac{1}{1 - \|z\|},$$

and therefore $G(z)$ is absolutely convergent. We extend the probability generating function to include time by defining:

$$G(z, t) = \sum_{i=0}^{\infty} p_i(t) z^i, \tag{7.8}$$

and use $G(z, t)$ to solve the birth and death model expressed in equation (7.6). Taking the derivative of $G(z, t)$ with respect to t and using (7.6) gives:

$$
\begin{aligned}
\frac{\partial G(z, t)}{\partial t} &= b \sum_n (n - 1) p_{n-1}(t) z^n \\
&\quad - (b + d) \sum_n n p_n z^n \\
&\quad + d \sum_n (n + 1) p_{n+1}(t) z^n \\
&= b \sum_n n p_n(t) z^{n+1} \\
&\quad - (b + d) \sum_n n p_n z^n \\
&\quad + d \sum_n n p_n(t) z^{n-1}.
\end{aligned}
\tag{7.9}
$$

Taking the derivative of $G(z, t)$ with respect to z gives:

$$\frac{\partial G(z, t)}{\partial z} = \sum_n n p_n(t) z^{n-1}. \tag{7.10}$$

Using (7.10) in (7.9), we get:

$$\frac{\partial G(z, t)}{\partial t} - (bz - d)(z - 1)\frac{\partial G(z, t)}{\partial z} = 0. \tag{7.11}$$

This partial differential equation can be solved using the method of characteristics [51]. To do so, we determine the initial and boundary conditions as follows:

1) Initial condition at $t = 0$. Suppose the population size at time $t = 0$ is n_0. Thus, $p_{n_0}(0) = 1$ and $p_n(0) = 0$ for $n \neq n_0$. In this case,

$$G(z, 0) = \sum_n p_n(0)z^n = z^{n_0}.$$

2) Boundary conditions at $z = 0$ and $z = 1$. These conditions are expressed by:

$$G(0, t) = p_0(t), \quad G(1, t) = \sum_n p_n(t) = 1.$$

The solution of (7.11) is constant along the characteristics (e.g., $G = z_0^n$), satisfying:

$$\frac{dz}{dt} = -(bz - d)(z - 1). \tag{7.12}$$

Using separation of variables, we find:

$$\left(\frac{bz - d}{z - 1} \right) e^{(b-d)t} = \frac{bz_0 - d}{z_0 - 1}.$$

Solving for z_0 gives:

$$z_0 = \frac{d(z - 1)e^{(b-d)t} - (bz - d)}{b(z - 1)e^{(b-d)t} - (bz - d)},$$

and therefore,

$$G(z, t) = \left(\frac{d(z - 1)e^{(b-d)t} - (bz - d)}{b(z - 1)e^{(b-d)t} - (bz - d)} \right)^{n_0}. \tag{7.13}$$

One can use (7.13) to determine the probability of extinction for the population, which is:

$$p_0(t) = G(0, t) = \left(\frac{d - de^{(b-d)t}}{d - be^{(b-d)t}} \right)^{n_0}.$$

Suppose $d > b$. Thus, $\lim_{t \to \infty} p_0(t) = 1$, and hence the population will go extinct. If $b > d$, then:

$$\lim_{t \to \infty} p_0(t) = \left(\frac{d}{b} \right)^{n_0} < 1.$$

This implies that even in the case $b > d$ the population can go extinct with nonzero probability. However, in the case of the deterministic model expressed by $n(t) = n_0 e^{(b-d)t}$, the population will never go extinct and grows unboundedly whenever $b > d$. This example highlights the difference between stochastic and deterministic models.

7.3   Markov Chains

Definition 7.3   A discrete-time stochastic process $\{X_n\}_{n=0}^{\infty}$ is said to have the Markov property if:

$$P(X_n = i_n | X_0 = i_0, \ldots, X_{n-1} = i_{n-1}) = P(X_n = i_n | X_{n-1} = i_{n-1}),$$

where $i_k \in \{1, 2, 3, \ldots\}$ and $k = 0, 1, 2, \ldots$. The stochastic process is then called a discrete-time *Markov chain*.

For a Markov chain, we define the probability mass function:

$$p_i(n) = P(X_n = i),$$

which means that the process is in state i at time n with probability $p_i(n)$. Now suppose the process is in state i at time n. Thus, at time $n + 1$, the process will be in state j, where j may be different from or the same as i. This represents a *one-step transition probability* defined as:

$$p_{ij}(n) = P(X_{n+1} = j | X_n = i).$$

☞ **Transition matrix.** The transition matrix of a discrete-time Markov chain $\{X_n\}_{n=1}^{\infty}$ is defined by:

$$T = \begin{pmatrix} p_{11} & p_{12} & p_{13} & \cdots \\ p_{21} & p_{22} & p_{23} & \cdots \\ \vdots & \vdots & \vdots & \ddots \\ p_{n1} & p_{n2} & p_{n3} & \cdots \\ \vdots & \vdots & \vdots & \ddots \end{pmatrix}.$$

For any fixed i th row in T, from the properties of PMF, we have $\sum_j p_{ij} = 1$. If the Markov chain is finite (i.e., $n \in \{1, 2, 3, \ldots, N\}$), then T is a square matrix of size $N \times N$.

Example 7.3   Suppose a mouse is released into a maze with four cells (representing states) as illustrated in Figure 7.2. We monitor the movements of the mouse in this maze, and make discrete observations every 2 minutes. We assume that the mouse is equally likely to stay in the same cell or move to another cell at each observation. Recording the movements of the mouse according to the observed cell numbers, we are interested in determining the one-step transition matrix for this Markov chain process.

Since we have four cells (or states), the Markov chain is finite with a transition matrix of size 4×4. Suppose that at the first observation the mouse is in state 1. At the next observation, the mouse could be either in state 1 with probability $p_{11} = 0.5$ or in state 2 with probability $p_{12} = 0.5$. At this observation, the mouse cannot be in state 3 or 4, as it requires more than one transition from state 1. Thus, $p_{13} = p_{14} = 0$.

Figure 7.2 Monitoring mouse movements in a maze.

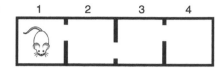

Now, suppose the mouse is in state 2 at the first observation. The probability that the mouse stays in the same state at the next observation is $p_{22} = 0.5$. The mouse cannot move to state 4 in one transition (as it requires two transitions), and therefore $p_{24} = 0$. Since the mouse is equally likely to go to any other states, it follows that $p_{21} = p_{23}$. Furthermore, $\sum_j^4 p_{2j} = 1$ implies that $p_{21} + p_{23} = 0.5$, and therefore $p_{21} = p_{23} = 0.25$.

A similar argument shows that if the mouse is in state 3 at the first observation, then we have $p_{33} = 0.5$, $p_{32} = p_{34} = 0.25$, and $p_{31} = 0$. If the mouse is in state 4 at the first observation, we can apply the same argument as when the mouse was in state 1, which gives $p_{44} = 0.5$, $p_{43} = 0.5$, and $p_{42} = p_{41} = 0$. Summarizing the above, we get the transition probability matrix:

$$T = \begin{bmatrix} 0.5 & 0.5 & 0 & 0 \\ 0.25 & 0.5 & 0.25 & 0 \\ 0 & 0.25 & 0.5 & 0.25 \\ 0 & 0 & 0.5 & 0.5 \end{bmatrix}.$$

Example 7.4 The birth and death model represents a Markov chain process. With the rates of b and d for birth and death respectively, we are interested in determining the transition matrix for the change in the population size during the time interval Δt (as the transition time between two states of the system). For any given i, representing the population size in the current state, the transition matrix can be defined by $T = [p_{ij}]$, where:

$$p_{ij} = \begin{cases} bi\Delta t + \mathcal{O}(\Delta t^2), & \text{for } j = i+1 \quad \text{(a birth occurs)}, \\ di\Delta t + \mathcal{O}(\Delta t^2), & \text{for } j = i-1 \quad \text{(a death occurs)}, \\ 1 - (b+d)i\Delta t + \mathcal{O}(\Delta t^2), & \text{for } i = j \quad \text{(no birth or death occurs)}, \\ \mathcal{O}(\Delta t^2), & \text{for } j \notin \{i, i-1, i+1\}. \end{cases}$$

☞ **Transition diagram.** A transition diagram represents the possible pathways from one state to any other state in a Markov chain process. Figure 7.3 shows the transition diagram for movements of the mouse discussed in Example 7.3.

Often we are interested in determining the probability of multi-step transitions from one state to another state in a Markov chain process. Considering the transition diagram in Figure 7.3, it is obvious that the probability of the mouse moving from state 1 to state 4 in two transitions is 0. However, the mouse can move from state 1 to state 3 in two transitions. This requires the mouse to move

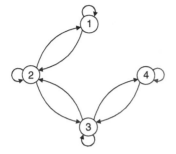

Figure 7.3 Transition diagram for the movements of the mouse in Example 7.3.

Figure 7.4 Representation of the areas for movements of a zebra.

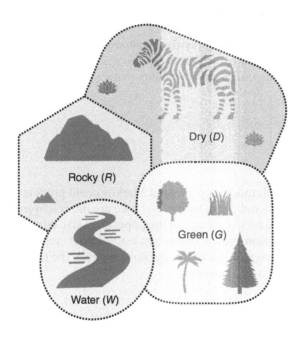

to state 2 in the first transition and move to state 3 in the second transition. Thus,

$$P(① \to ③ \text{ in } 2 \text{ transitions}) = p_{12}p_{23} = 0.5 \times 0.25 = 0.125.$$

Example 7.5 A zebra moves among areas of different types as represented in Figure 7.4. Suppose that staying in or moving to the green area is twice as likely as staying in or moving to any other area, but movement between other areas is equally likely. One can follow the same process as in the example of the mouse and maze to determine the probabilities that the zebra moves from one area to another between two consecutive observations.

Figure 7.5 Transition diagram for the movements of the zebra in Example 7.5.

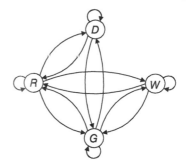

For example, if the zebra is in state D at the first observation, then the probability of moving to state W in one transition is $p_{D \to W} = 0$. The probability of moving to state G is twice larger than the probability of staying in state D or moving to state R. Thus,

$$p_{D \to G} = 2p_{D \to D} = 2p_{D \to R}.$$

From the relation:

$$p_{D \to D} + p_{D \to G} + p_{n \to R} + p_{n \to W} = 1,$$

we find that:

$$p_{D \to D} = \frac{1}{4}, \quad p_{D \to G} = \frac{1}{2}, \quad p_{D \to R} = \frac{1}{4}, \quad p_{D \to W} = 0.$$

One can find the transition probabilities for other states in a similar manner. Hence, the one-step transition matrix for the movements of the zebra is:

$$T = \begin{array}{c} \\ \\ \end{array} \begin{array}{cccc} D & R & G & W \\ \left[\begin{array}{cccc} \frac{1}{4} & \frac{1}{4} & \frac{1}{2} & 0 \\ \frac{1}{5} & \frac{1}{5} & \frac{2}{5} & \frac{1}{5} \\ \frac{1}{9} & \frac{1}{9} & \frac{2}{3} & \frac{1}{9} \\ 0 & \frac{1}{4} & \frac{1}{2} & \frac{1}{4} \end{array} \right] & \begin{array}{c} D \\ R \\ G \\ W \end{array} \end{array}$$

A transition diagram corresponding to this one-step transition matrix is shown in Figure 7.5. Suppose we are interested in determining the transition probability for the zebra going from state D to state W in exactly two steps. Figure 7.6 shows the possible pathways for this two-step transition. Using these pathways, we have:

$$P\left(\text{(D)} \longrightarrow \text{(W)} \text{ in 2 transitions} \right) = P\left(\text{(D)} \overset{1}{\to} \text{(R)} \text{ and } \text{(R)} \overset{2}{\to} \text{(W)} \right)$$

$$+ P\left(\text{(D)} \overset{1}{\to} \text{(G)} \text{ and } \text{(G)} \overset{2}{\to} \text{(W)} \right)$$

$$= \frac{1}{4} \times \frac{1}{5} + \frac{1}{2} \times \frac{1}{9}$$

$$= \frac{1}{20} + \frac{1}{18} = \frac{19}{180}.$$

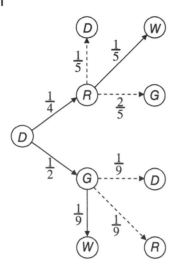

Figure 7.6 Possible pathways for the zebra moving from the dry area (*D*) to water (*W*) in exactly two transitions (Figure 7.5). The probabilities of moves between states are indicated on the connecting arrows.

Theorem 7.1 (Multi-step transitions) Let $T = [p_{ij}]$ be the one-step transition matrix of a Markov chain process. The (i, j) th entry of the m-step transition matrix is the (i, j) th entry of T^m.

Using this theorem, we can now find the probability that the mouse in Example 7.3 moves from one state to any other state in exactly two steps. The two-step transition matrix is:

$$
T^2 = \begin{bmatrix} 0.5 & 0.5 & 0 & 0 \\ 0.25 & 0.5 & 0.25 & 0 \\ 0 & 0.25 & 0.5 & 0.25 \\ 0 & 0 & 0.5 & 0.5 \end{bmatrix} \times \begin{bmatrix} 0.5 & 0.5 & 0 & 0 \\ 0.25 & 0.5 & 0.25 & 0 \\ 0 & 0.25 & 0.5 & 0.25 \\ 0 & 0 & 0.5 & 0.5 \end{bmatrix}
$$

$$
= \begin{bmatrix} 0.375 & 0.5 & 0.125 & 0 \\ 0.25 & 0.4375 & 0.25 & 0.0625 \\ 0.0625 & 0.25 & 0.4375 & 0.25 \\ 0 & 0.125 & 0.5 & 0.375 \end{bmatrix}.
$$

The two-step transition matrix for the movements of the zebra in Example 7.4 is:

$$
T^2 = \begin{bmatrix} \frac{1}{4} & \frac{1}{4} & \frac{1}{2} & 0 \\ \frac{1}{5} & \frac{1}{5} & \frac{2}{5} & \frac{1}{5} \\ \frac{1}{9} & \frac{1}{9} & \frac{2}{3} & \frac{1}{9} \\ 0 & \frac{1}{4} & \frac{1}{2} & \frac{1}{4} \end{bmatrix} \times \begin{bmatrix} \frac{1}{4} & \frac{1}{4} & \frac{1}{2} & 0 \\ \frac{1}{5} & \frac{1}{5} & \frac{2}{5} & \frac{1}{5} \\ \frac{1}{9} & \frac{1}{9} & \frac{2}{3} & \frac{1}{9} \\ 0 & \frac{1}{4} & \frac{1}{2} & \frac{1}{4} \end{bmatrix}
$$

$$= \begin{bmatrix} \dfrac{121}{720} & \dfrac{121}{720} & \dfrac{67}{120} & \dfrac{19}{180} \\[6pt] \dfrac{121}{900} & \dfrac{83}{450} & \dfrac{41}{75} & \dfrac{121}{900} \\[6pt] \dfrac{67}{540} & \dfrac{41}{270} & \dfrac{3}{5} & \dfrac{67}{540} \\[6pt] \dfrac{19}{180} & \dfrac{121}{720} & \dfrac{67}{120} & \dfrac{121}{720} \end{bmatrix}.$$

Example 7.6 Consider the transition diagram in Figure 7.7. We are interested in knowing the probability of ever reaching state C, given that we start from state A. The first possibility for going from A to C is to follow the path $A \to B \to C$. This path has the probability:

$$p_1 = P_{A \to B} P_{B \to C} = \frac{1}{2} \times \frac{3}{9} = \frac{3}{18}.$$

The second possibility is to follow the path $A \to B \to A \to B \to C$. This path has the probability:

$$p_2 = P_{A \to B} P_{B \to A} P_{A \to B} P_{B \to C} = \frac{1}{2} \times \frac{4}{9} \times \frac{1}{2} \times \frac{3}{9} = \frac{3}{18} \times \frac{2}{9}.$$

The third possibility is to follow the path $A \to B \to A \to B \to A \to B \to C$. This path has the probability:

$$p_3 = P_{A \to B} P_{B \to A} P_{A \to B} P_{B \to A} P_{A \to B} P_{B \to C} = \frac{1}{2} \times \frac{4}{9} \times \frac{1}{2} \times \frac{4}{9} \times \frac{1}{2} \times \frac{3}{9}$$
$$= \frac{3}{18} \left(\frac{2}{9}\right)^2.$$

Continuing these pathways, one can calculate the probability of ever reaching state C from state A as:

$$p_{A \to C} = p_1 + p_2 + p_3 + \cdots = \frac{3}{18} + \frac{3}{18} \times \frac{2}{9} + \frac{3}{18} \left(\frac{2}{9}\right)^2 + \cdots$$
$$= \left(\frac{3}{18}\right) \left[1 + \frac{2}{9} + \left(\frac{2}{9}\right)^2 + \cdots\right]$$
$$= \frac{3}{18} \sum_{n=0}^{\infty} \left(\frac{2}{9}\right)^n$$
$$= \frac{3}{18} \times \frac{1}{1 - \frac{2}{9}} = \frac{3}{14}.$$

Figure 7.7 Transition diagram for a Markov chain process.

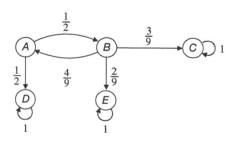

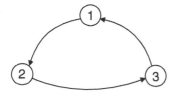

Figure 7.8 Transition diagram for a Markov chain process.

Example 7.7 Suppose we have a Markov chain process as represented in Figure 7.8. This diagram indicates that $p_{31} = p_{12} = p_{23} = 1$, and any other $p_{ij} = 0$ in one-step transitions. Thus the one-step transition matrix is:

$$T = \begin{bmatrix} 0 & 1 & 0 \\ 0 & 0 & 1 \\ 1 & 0 & 0 \end{bmatrix}.$$

Using the theorem for multi-step transitions, we find that:

$$T^3 = \begin{bmatrix} 1 & 0 & 0 \\ 0 & 1 & 0 \\ 0 & 0 & 1 \end{bmatrix}.$$

and therefore $T^m = T^3$ for $m = 3k$. Thus, in this Markov chain process, each state can reach itself in three transitions. Furthermore, each state can be reached from any other state in less than 3 transitions. This brings us to the concept of accessibility.

☞ **Accessibility.** Suppose we have a discrete-time Markov chain process. We say state j is *accessible* from state i if $p_{ij}(k) > 0$ for some $k \geq 1$. We say states i and j are *mutually accessible* if state i is accessible from state j, and state j is accessible from state i.

From the transition diagram in Figure 7.8, it can be seen that each pair of states is mutually accessible. However, for the transition diagram in Figure 7.9, it is impossible to reach states 1, 3, or 4 from states 2 and 5. For these types of Markov chain process, we can break down the system into sets of mutually accessible states, where there is no common state in any pair of sets. For example, for the transition diagram in Figure 7.9, we may consider the following sets:

$$S_1 = \{1, 3, 4\}, \quad S_2 = \{2, 5\}.$$

Figure 7.9 Transition diagram for a Markov chain process.

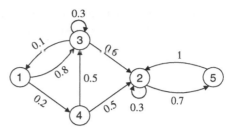

Using these sets, we can reorder the one-step transition matrix to obtain:

$$
T =
\begin{array}{c}
S_2 \left\{ \begin{array}{c} \\ \\ \end{array} \right. \\
\\
S_1 \left\{ \begin{array}{c} \\ \\ \\ \end{array} \right.
\end{array}
\overbrace{}^{S_2} \overbrace{}^{S_1}
\left[
\begin{array}{cc|ccc}
0.3 & 0.7 & 0 & 0 & 0 \\
1 & 0 & 0 & 0 & 0 \\
\hline
0 & 0 & 0 & 0.8 & 0.2 \\
0.6 & 0 & 0.1 & 0.3 & 0 \\
0.5 & 0 & 0 & 0.5 & 0
\end{array}
\right] ,
$$

where the 2×2 block of entries correspond to the transition probabilities between states in set S_2, and the 3×3 block of entries correspond to the transition probabilities between states in set S_1. In this form, T is a lower block-triangular matrix, and therefore T^m remains a lower block-triangular matrix for $m \geq 1$. It is also possible to reorder the transition matrix to obtain an upper block-triangular matrix:

$$
T =
\begin{array}{c}
S_1 \left\{ \begin{array}{c} \\ \\ \\ \end{array} \right. \\
\\
S_2 \left\{ \begin{array}{c} \\ \\ \end{array} \right.
\end{array}
\overbrace{}^{S_1} \overbrace{}^{S_2}
\left[
\begin{array}{ccc|cc}
0 & 0.8 & 0.2 & 0 & 0 \\
0.1 & 0.3 & 0 & 0.6 & 0 \\
0 & 0.5 & 0 & 0.5 & 0 \\
\hline
0 & 0 & 0 & 0.3 & 0.7 \\
0 & 0 & 0 & 1 & 0
\end{array}
\right] .
$$

☞ **Canonical form.** Suppose $S_1, S_2, \ldots, S_n$ is a family of mutually accessible sets for a Markov chain process. Then a canonical form of the transition matrix is given by:

$$
T =
\begin{pmatrix}
A_1 & 0 & \cdots & 0 \\
T_{21} & A_2 & \cdots & 0 \\
\vdots & \vdots & \ddots & \vdots \\
T_{n,1} & \cdots & T_{n,n-1} & A_n
\end{pmatrix} ,
$$

where A_i is the transition matrix between states in S_i, and T_{ij} ($j < i$) is the transition matrix between states in S_i and S_j. If the one-step transition matrix T has a canonical form, then T^m also has a canonical form for $m > 1$.

7.4 Random Walks

Definition 7.4 Let $\{X_n\}_{i=1}^{\infty}$ be a stochastic process as a sequence of independent discrete random variables. For each positive integer n, we let S_n denote the sum $\sum_{i=1}^{n} X_i$. The sequence $\{S_n\}_{n=1}^{\infty}$ is called a *random walk*.

Since the X_i are independent random variables, the probability of any particular sequence of outcomes can be obtained by multiplying the probabilities that each X_i takes on the specified value in the sequence. The simplest random walk corresponds to the experiment of flipping a coin. In this case, the probability distribution function for the random variables X_i is given by:

$$F(x) = \begin{cases} \frac{1}{2}, & \text{if } x = \pm 1, \\ 0, & \text{otherwise,} \end{cases}$$

where $x = \pm 1$ correspond to the outcomes heads and tails. In this random walk, all paths of length n have the same probability 2^{-n}.

Example 7.8 (The drunkard's walk) Suppose a drunkard walks the street with a bar at the beginning. Suppose the drunkard moves in discrete units of one step either to the left or to the right. We are interested in determining the probability that the drunkard is at position A after n steps. We assume that the drunkard's walk follows the Bernoulli process with probability p of going one step to the left. Thus, the probability of going one step to the right is $q = 1 - p$. Suppose that the i th step in this random walk is represented by the random variable X_i, taking the values -1 for a step to the left and $+1$ for a step to the right. Thus,

$$S_n = X_1 + X_2 + \cdots + X_n,$$

gives the position of the drunkard after n steps, and therefore $S_n = A$. Let a and b represent the number of steps taken in the right and left directions, respectively. Thus, $n = a + b$, $A = a - b$, and we have:

$$a = \frac{1}{2}(A + n), \quad b = \frac{1}{2}(n - A).$$

The number of possible ways to take b steps to the left among a total of n steps is:

$$\text{number of ways to take } b \text{ steps to the left} = \binom{n}{b},$$

which is also the number of ways to arrive at position A. Each way has the probability $p^a q^b$, and therefore,

$$P(S_n = A) = \begin{cases} \binom{n}{b} p^a q^b, & \text{if } A = n - 2b, \\ 0, & \text{otherwise.} \end{cases}$$

In the special case where $p = q = 1/2$, we have:

$$P(S_n = A) = \begin{cases} \binom{n}{b} \frac{1}{2^n}, & \text{if } A = n - 2b, \\ 0, & \text{otherwise.} \end{cases}$$

☞ **Return in random walks.** Suppose $\{S_n\}_{n=1}^{\infty}$ is a random walk. We say that a *return* to the origin has occurred at time m if $S_m = 0$.

Example 7.9   Consider the example of a random walk by a drunkard. The probability that the drunkard returns to the bar (at the beginning of the street; i.e., $A = 0$ or $a = b$) after walking $2n$ steps is:

$$P(S_{2n} = 0) = \binom{2n}{n} p^n q^n.$$

In the special case where $p = q = 1/2$, this probability is:

$$P(S_{2n} = 0) = \binom{2n}{n} 2^{-2n}.$$

Exercises

1   Suppose one event has occurred during the transition time in the birth and death model. What is the probability that this event was a birth? What is the probability that this event was a death?

2   Determine the transition probabilities for the Markov birth and death process.

3   Suppose that the growth of a population follows the logistic equation, given by:

$$N' = rN\left(1 - \frac{N}{K}\right),$$

where r is the growth rate and K is the carrying capacity of the population. Develop a stochastic Markov chain model corresponding to this deterministic representation.

4 A computer has two operating systems. Without interference, the computer either continues to work with the same operating system or switches to the other one every hour. The probability of switching between operating systems is given by the transition probability matrix:

$$T = \begin{bmatrix} 0.48 & 0.52 \\ 0.6 & 0.4 \end{bmatrix}.$$

a) If the computer is on with the first operating system at 2:30, what is the probability that the computer is on with the same operating system at 4:30?

b) If the computer is on with the first operating system at 2:30, what is the probability that the computer is on with the second operating system at 5:30?

5 A mouse is released into a maze as shown in the figure below. Observations are made every 2 minutes. At any observation, the mouse is in one of the states 1, ... , 7. Suppose that the probability of being in the same state in two consecutive observations is the same as the probability of being in a different state. Find the one-step transition probability matrix associated with the Markov process of these movements.

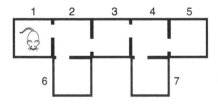

6 Suppose that a Markov chain has a transition diagram as shown below, with the transition probabilities given on each path.

a) Determine the canonical form of its one-step transition matrix.

b) What is the probability of moving from state 2 to state 3 in three steps? Identify all the pathways for this three-step transition.

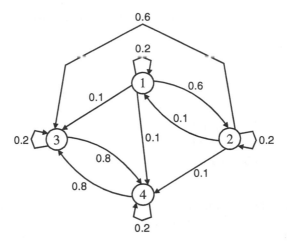

7 Consider a random walk between 4 points as shown in the figure below with an immediate reflection at the end points (dashed arrows).

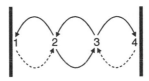

Suppose we have the following transition probabilities:

$$p_{i,i+1} = \frac{1}{1+\alpha},$$
$$p_{i,i-1} = \frac{\alpha}{1+\alpha}.$$

a) Find the one-step transition probability matrix for this Markov chain process.
b) Using this matrix, show that reaching state 2 in two transitions is possible only from states 4 and 2.

8 Consider a drunkard who is on a bridge and is attempting to cross it. Assume that the bridge is 5 feet wide. Let X_n be the distance, measured in 1-foot segments, from the left edge of the bridge to the drunkard at step n. The sample space of the chain $\{X_n\}_{n=0,1,2,...}$ is thus $S = \{0, 1, 2, 3, 4, 5\}$. At any time, if not at an edge of the bridge, the drunkard can sidestep 1 foot to the right with probability 0.4 or 1 foot to the left with probability 0.6. If at an edge ($X_n = 0, X_n = 5$), the drunkard steps back to the previous position. Find the one-step transition matrix for this Markov process.

9 Consider the following transition matrices. Find P^n for all $n > 1$.

a)

$$P = \begin{bmatrix} 0 & 0.5 & 0.5 \\ 1 & 0 & 0 \\ 1 & 0 & 0 \end{bmatrix}.$$

b)

$$P = \begin{bmatrix} 1 & 0 & 0 \\ 1 & 0 & 0 \\ 0.5 & 0.5 & 0 \end{bmatrix}.$$

10 Write the following transition diagram as a transition matrix. Then calculate the two-step transition matrix and draw its transition diagram.

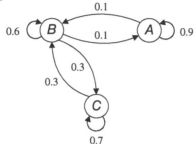

11 Using a transition diagram, calculate the probability that the white knight on the chess board goes from its current location to the location highlighted by the star in at most three moves.

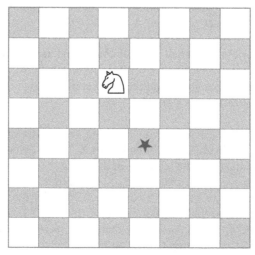

12 Consider a Markov chain process with the transition diagram shown in the figure below.

a) If we start in state 1, what is the probability that we are still in state 1 after three steps? After five steps? After one thousand steps?

b) If we start in state 4, what is the probability that we ever reach state 7?

c) If we start in state 3, what is the probability that we ever reach state 8?

d) Identify all the pathways by which we can reach from state 4 to state 8 in five transitions. Calculate the probability for these transitions.

e) Write the canonical form of the transition matrix for this Markov chain, and identify the sets of mutually accessible states.

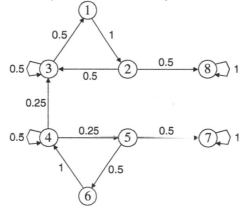

8

Computer Simulations

Computer simulations often rely on discrete methods to numerically approximate the solutions of a continuous model. In this chapter, we discuss the implementation of discrete methods presented in Chapter 5 for deterministic models. We also discuss the stochastic implementation of such models using the Gillespie direct algorithm [55]. For the purpose of simulations, we will provide examples of computer codes that are written in Matlab© software (developed by The MathWorks Inc.) to illustrate the implementation of discretization methods and their results. We use Matlab© software for a number of reasons. It has a user-friendly graphical interface with important features and has been widely used for computational and simulation purposes. Matlab© provides a convenient environment to perform various types of calculations, in particular for solving differential equations. As a *matrix laboratory*, Matlab© uses matrices as a basic data structure to perform calculations. Since discrete methods rely on iterative calculation of vectors, matrix operations play an important role in the computer implementation of such methods.

8.1 Deterministic Structure

To implement discrete methods for deterministic models, we need to create a vector that discretizes the time interval. For a fixed time-step h, a time vector (Tvec) of length $N = (b - a)/h$ can be generated by using the command line:

```
>> Tvec=a:h:b;
```

We will also need to generate vectors of the same length as Tvec for each variable of the system to update them as approximations are made at each time-step. Initially, one can create these vectors with zero entries. The following command line generates a matrix of size $m \times n$ with zero entries:

```
>> zeros(m,n);
```

Mathematical Modelling: A Graduate Textbook, First Edition. Seyed M. Moghadas and Majid Jaberi-Douraki.
© 2019 John Wiley & Sons, Inc. Published 2019 by John Wiley & Sons, Inc.
Companion Website: www.wiley.com/go/Moghadas/Mathematicalmodelling

In order to create a vector for a system variable, say X, we may use the following command lines:

```
>> L=length(Tvec);
>> X=zeros(L,1);
```

The command `zeros(L,1)` generates a vector of length L, which is the same as a matrix of size $L \times 1$. The elements of this vector correspond to the time-steps that will be stored in `Tvec`, where `X(1)` is the initial condition given to the system at time `t=a`. During the simulations, successive entries of this vector are updated as numerical calculations are made at each time-step. Once the program is compiled and completed, the numerical results can be accessed from this vector. For example, one can plot the variable X as a function of time by using the command line:

```
>> plot(Tvec,X);
```

For writing a computer code, some basic knowledge of the programming language is necessary. Since we use Matlab© for computational purposes, the Matlab© Primer [27] provides useful information for developing simulation algorithms.

Example 8.1 Suppose we are interested in the time profiles of the prey and predator populations given by the system:

$$x' = rx(1 - x) - axy,$$
$$y' = cxy - dy, \tag{8.1}$$

where the parameters $r, a, c,$ and d are positive constants. Computer code for simulating the variables x and y using the Euler method is provided in Listing 8.1. Figure 8.1 shows the time profiles of the x and y variables, with damped oscillations towards a stable point. This indicates that there is a stable critical point in the $(x > 0, y > 0)$ region. Figure 8.2 shows that the solution with initial condition $(x_0, y_0) = (0.1, 0.1)$ approaches this stable point through a spiral curve. In these simulations, the parameter values are $r = 0.1$, $a = 6$, $c = 5$, and $d = 1.5$. The time interval is $[0, 500]$ and the time-step is fixed at $h = 0.001$.

Listing 8.1: Matlab© code for simulating the predator–prey system (8.1). To run the simulations, parameter values, initial conditions, time interval, and time-step must be given as input.

```
1  %% Parameters of the system
2  r=r0;     % r0 is growth rate of prey
3  a=a0;     % a0 is predation rate
4  c=c0;     % c0 is conversion rate
```

```
 5  d=d0;     % d0 is natural death rate of predator
 6  %% Creating a vector of time for discritization
 7  t_start=T0;     % T0 is the beginning of time interval
 8  t_end=Tn;       % Tn is the end of time interval
 9  h=h0;           % h0 is a fixed step-size for discretiziation
10  t=T0:h:Tn;      % creating the time vector
11  Lt=length(t);   % Lt is the length of time vector
12  %% Creating vectors of system variables
13  x=zeros(Lt,1);  % creates a vector of Lt by 1 zeros for x
14  y=zeros(Lt,1);  % creates a vector of Lt by 1 zeros for y
15  %% Initial conditions
16  x(1)=x0;        % x0 is the initial value of x at time T0
17  y(1)=y0;        % y0 is the initial value of y at time T0
18  %% Euler method
19  for j=1:Lt
20      x(j)=x(j-1)+h*(r*x(j-1)*(1-x(j-1))-a*x(j-1)*y(j-1));
21      y(j)=y(j-1)+h*(c*x(j-1)-d)*y(j-1);
22  end
23  %% Plotting the variables (solutions) as a function of time
24  plot(t,x);        % plotting x
25  figure            % new figure
26  plot(t,y);        % plotting y
27  figure            % new figure
28  plot(t,[x,y]);    % plotting x and y on the same figure
29  figure            % new figure
30  plot(x,y);        % plotting phase plane
```

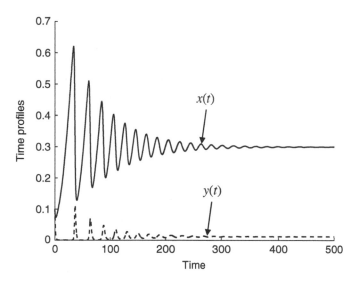

Figure 8.1 Time profiles of the prey (solid line) and predator (dashed line) populations simulated with the Euler method using parameter values $r = 0.1$, $a = 6$, $c = 5$, and $d = 1.5$ in the time interval [0, 500] with fixed time-step $h = 0.001$.

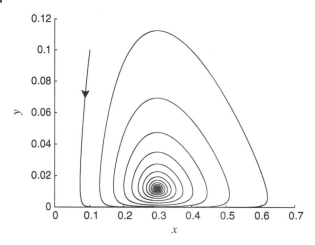

Figure 8.2 Phase plane for the predator–prey system (8.1) simulated with the Euler method using parameter values $r = 0.1, a = 6, c = 5$, and $d = 1.5$ in the time interval [0, 500] with fixed time-step $h = 0.001$.

8.2 Stochastic Structure

Since stochastic models involve randomness, the occurrence of events depends on the rates of movement between different compartments. The transitions between compartments are certain to happen in deterministic models. However, in stochastic models, such transitions are associated with probabilities as discussed in Chapter 7. There are a number of well-established methods to simulate stochastic models. Here, we restrict our attention to the Gillespie direct algorithm [55], which is widely used in stochastic processes.

Let t be a continuous variable, and define the random vector $X(t)$ of the state variables of the system at time t, for $t \in [0, \infty)$. Suppose Δt is the transition period during which an event may occur. We define $\Delta X(t) = X(t + \Delta t) - X(t)$, which represents changes that occur to the random vector X during Δt units of time. We now define the transition probability as:

$$P(\Delta X(t) = \Theta(\text{state variable}) \mid X(t)),$$

where,

$$\Theta(\text{state variable}) = \begin{cases} -1, & \text{decrease in the state variable,} \\ 0, & \text{no change in the state variable,} \\ +1, & \text{increase in the state variable.} \end{cases}$$

We may assume that Δt is sufficiently small, so that at most one change of state can occur during the transition time Δt. To use the Gillespie direct algorithm, transition rates in the continuous model are converted to transition probabilities to determine which event takes place and what changes occur to the state variables. Suppose that a_i is the transition rate associated with event i in the model. Then the probability of event i happening is obtained as:

$$P(\text{Event } i) = \frac{a_i}{\sum_i a_i}.$$

In this formulation, the time to the next event (Δt) is exponentially distributed with the parameter equal to the sum of the rates for all possible events. The probability density function is given by:

$$f(\Delta t) = \left(\sum_i a_i \right) e^{-\Delta t \sum_i a_i}. \tag{8.2}$$

Using inverse transform sampling [10], one can estimate the time to the next event. For a given random number r drawn from the uniform distribution on the unit interval $(0, 1]$, Δt can be estimated as $- \ln(r)/\sum_i a_i$, which is obtained from equation (8.2). This shows that, depending on the state of the system, the transition time for an event to occur may change. Once the transition time is determined, one can order the events as an increasing fraction of $\sum_i a_i$ and generate another random number between 0 and 1 to determine the nature of the next event.

In order to better understand the Gillespie direct algorithm, we consider the simple example of an SIR epidemic model.

Example 8.2 Suppose we have an epidemic with state variables S, I, and R, respectively, representing the population sizes of susceptible, infected, and recovered individuals at any time. Let β and γ be the transmission and recovery rates of infection, respectively. The deterministic structure of the system is:

$$S' = -\beta SI,$$
$$I' = \beta SI - \gamma I,$$
$$R' = \gamma I. \tag{8.3}$$

In this model, the transition from S to I occurs as a result of an event associated with disease transmission at the rate βSI, referred to as incidence. Thus the transition probability with Δt units of time is $\beta SI \Delta t + \mathcal{O}(\Delta t^2)$. Similarly, for the event of recovery that is associated with a transition from I to R, we have the transition probability $\gamma I \Delta t + \mathcal{O}(\Delta t^2)$. In the Gillespie direct algorithm, the states of the system are updated according to the events that occur during the transition time Δt. These states can be summarized through the following transitions and their associated probabilities.

Event	Transition	Transition probability
infection of a susceptible individual	$S \to S-1$	$\beta SI \Delta t + \mathcal{O}(\Delta t^2)$
	$I \to I+1$	
recovery of an infected individual	$I \to I-1$	$\gamma I \Delta t + \mathcal{O}(\Delta t^2)$
	$R \to R+1$	

We now proceed with stochastic simulations using the Gillespie direct algorithm. For this implementation, we use `function` in Matlab© to create different modules that have specific inputs and outputs, and can be called during the simulation process.

Example 8.3 Consider the SIR epidemic model in (8.3). In this example, we will simulate the system using the Gillespie direct algorithm. We show the simulation logic in Figure 8.3.

Based on the logic diagram, we write the algorithm in four separate modules (Listings 8.2–8.5). The first three modules are functions. When the last module is run, the simulations start and call the third module which is the simulation function. In the third module, the function for stochastic iterations from the second module is called. And finally, in the second module, the function for updating the state variables and their changes according to the transition probabilities is called from the first module. This modularization allows us to monitor the algorithm closely and resolve the issues when coding errors emerge.

Listing 8.2: Matlab© code for the SIR epidemic model, part I. Module for updating the state variables.

```
1  function[step, new_value]=Update_Variables(old, Parameters)
2  % Parameters of the model
3  beta=Parameters(1);
4  gamma=Parameters(2);
5  % Variables of the model
6  S=old(1);
7  I=old(2);
8  R=old(3);
9  Change=zeros(2,3);   % update the variables with events
10 %% Occurrence of infection
11 Rate(1) = beta*S*I;
12 Change(1,:)= [-1  +1  0]*sign(heaviside(S-1)); % S-->I (if S>=1)
13 %% Recovery from infection
14 Rate(2)  = gamma*I;
15 Change(2,:)= [0  -1  +1]*sign(heaviside(I-1)); % I-->R (if I>=1)
16 %% Update rates
17 Random1=rand(1,1);
18 Random2=rand(1,1);
```

```
19  % Estimate the transition time
20  if(sum(Rate) > 0)
21      step = -log(Random1)/(sum(Rate));
22  else
23      step=0.1;
24      new_value=old;
25      return
26  end
27  % Find which event occurs
28  m=min(find(cumsum(Rate)>=Random2*sum(Rate)));
29  new_value=old+Change(m,:);
```

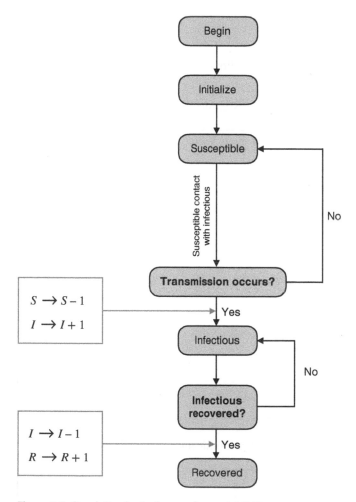

Figure 8.3 Simulation logic diagram for model (8.3).

Listing 8.2 provides the algorithm for introducing parameters and variables, and updating the variables based on whether an infection or recovery event occurs within the transition time estimated by:

```
step=-log(Random1)/(sum(Rate));
```

In order to update the states S, I, and R, we use a combination of Heaviside and sign functions. Since the change in a population state can occur only when the population is positive, we consider `sign(heaviside(Pop-1))` to allow for the change with a unit of population (i.e., one individual). The Heaviside function is defined by:

$$H(x) = \begin{cases} 0, & \text{if } x < 0, \\ \frac{1}{2}, & \text{if } x = 0, \\ 1, & \text{if } x > 0. \end{cases}$$

Thus, `sign(heaviside(Pop-1))` allows for a change in the population size to occur only if Pop ≥ 1.

Listing 8.3 : Matlab$^{\copyright}$ code for the SIR-epidemic model, part II. Module for stochastic iterations.

```
1  function [T,P]=Stoch_Iteration(Time,Initial,Parameters)
2  S=Initial(1);
3  I=Initial(2);
4  R=Initial(3);
5  % Initiation of iterations
6  T=0; P(1,:)=[S I R];
7  old=[S I R];
8  loop=1;
9  while (T(loop)<Time(2))
10     [step,new]=Update_Variables(old,Parameters);
11     loop=loop+1;
12     T(loop)=T(loop-1)+step;
13     P(loop,:)=old;
14     loop=loop+1;
15     T(loop)=T(loop-1);
16     P(loop,:)=new;
17     old=new;
18     if loop>=length(T)
19         T(loop*2)=0;
20         P(loop*2,:)=0;
21     end
22  end
```

Listing 8.4: Matlab$^{\copyright}$ code for the SIR epidemic model, part III. Module for simulations.

```
1  function [t,S,I,R] = Simulations(beta,gamma,S0,I0,R0,tmax)
2  % Main iterations
```

```
3  [t, pop] = Stoch_Iteration([0 tmax],[S0,I0,R0],[beta,gamma]);
4  S=pop(:,1);
5  I=pop(:,2);
6  R=pop(:,3);
```

For the parameter values given in Listing 8.5, the program was compiled using two CPU cores to run 100 simulations in each core. Figure 8.4 shows the outputs for the state variable I (i.e., the number of infections at any point in time) for 200 independent realizations (gray curves), and the average of these realizations (black curve). As is evident, each realization produces a different infection curve as a result of stochasticity. In contrast, one can see that using the same parameters and initial conditions, a deterministic method will generate the same curve every time the program is run. This can be easily studied by developing an Euler method or a nonstandard finite difference method, and we leave it as an exercise.

Listing 8.5: Matlab© code for the SIR epidemic model, part IV. Module for inputs and averaging outputs.

```
1   %% Initial populations
2   S0=1000;
3   I0=1;
4   R0=0;
5   %% Parameters of the model
6   R0=2;                % reproduction number
7   gamma=1/4;           % recovery rate (per day)
8   beta=gamma*R0/S0;    % transmission rate (per day per person)
9   %% Time of simulations
10  tmax=100;            % number of days
11  ttime=0:1:tmax;      % daily step-size for averaging
12  nsims=200;           % number of realizations
13  %% Main simulations
14  ll=0;
15  for ii=1:nsims
16      ncore=2;         % number of computer cores for parallelization
17      parfor i=1:ncore
18          [t{i,ii},S{i,ii},I{i,ii},R{i,ii}] = Simulations(beta,...
19              gamma,S0,I0,R0,tmax);
20      end
21      for i=1:ncore
22          ll=ll+1;
23          SS(ll,1)=S{i,ii}(1);
24          II(ll,1)=I{i,ii}(1);
25          RR(ll,1)=R{i,ii}(1);
26      end
27      k=2;
28      for j=2:length(t{i,ii})
29          for bb=k:length(ttime)
30              if ttime(bb)<=t{i,ii}(j)
31                  SS(ll,bb)=S{i,ii}(j);
32                  II(ll,bb)=I{i,ii}(j);
33                  RR(ll,bb)=R{i,ii}(j);
```

```
34                     else
35                          break
36                     end
37              end
38              k=bb;
39        end
40  end
41  %% Averaging all the simulations for each variable
42  AveS=sum(SS,1)/size(SS,1);
43  AveI=sum(II,1)/size(II,1);
44  AveR=sum(RR,1)/size(RR,1);
```

It is also interesting to note that the average of realizations has a lower magnitude compared to some realizations. This is due to the fact that in many simulations, the initially infectious individuals ($I(0)$) recover without infecting any susceptible individuals and therefore the epidemic dies out. As we have seen in Chapter 1, Example 1.10, infection is initially expected to grow in the population if $R_0 > 1$. In our simulations here, $R_0 = \beta S_0/\gamma = 2$, and therefore in a deterministic model an epidemic occurs. However, in a stochastic model, the epidemic may still die out due to random effects even when $R_0 > 1$. Using the analysis of the birth and death model in Chapter 7, one can determine the probability of disease propagation (i.e., epidemic). Recall that the probability of population extinction is:

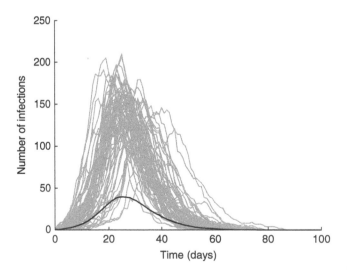

Figure 8.4 Stochastic simulations of the SIR epidemic model with $\beta = 5 \times 10^{-4}$ and $\gamma = 1/4$ in a population with initial conditions $S(0) = 1000$, $I(0) = 1$, and $R(0) = 0$. Gray curves show 200 independent realizations, and the black curve represents the average of these realizations. Simulations were performed using Listings 8.2–8.5.

$$P(\text{extinction}) = \begin{cases} \left(\frac{d}{b}\right)^{n_0}, & \text{if } b > d, \\ 1, & \text{if } b \leq d, \end{cases}$$

where n_0 is the initial population size, and b and d are the birth and death rates, respectively. In the context of our SIR model discussed here, we can consider the occurrence of infection as a birth in the infected population, and the event of recovery as a death. From the notation used in Example 1.10, we find that $d/b = 1/R_0$ in the epidemic model. Thus, the probability of no epidemic is:

$$P(\text{no epidemic}) = \begin{cases} \left(\frac{1}{R_0}\right)^{I(0)}, & \text{if } R_0 > 1, \\ 1, & \text{if } R_0 \leq 1. \end{cases} \tag{8.4}$$

Given this probability, and using the parameter values in Listing 8.5, we expect about $(1/2)^{(I_0=1)} = 0.5$ (or 50%) of the simulations in the stochastic model not to result in an epidemic.

Example 8.4 Let us consider the random walk process for the drunkard in Chapter 7. We assume that the drunkard will go one step to the right or to the left with probability $p = 0.5$. One can therefore develop a stochastic simulation for this process by defining the random variable at each step, and updating the values of this variable by $+1$ or -1, depending on whether the step is to the right or to the left. Listing 8.6 provides simple Matlab© code to simulate this random walk. Initiating the random variable at the origin, Figure 8.5 shows the results of this random walk for several runs.

Listing 8.6: Matlab© code for the drunkard's random walk.

```
1  p=0.5;      % probability of going to the left or right
2  steps=0:100;        % number of steps
3  Lt=length(steps);   % length of steps vector
4  walk=zeros(1,Lt);   % creating the vector of walks
5  walk(1)=0;          % initiating walks at the origin
6  for i=2:Lt
7      if rand<p       % sampling a random number in [0, 1]
8          walk(i)=walk(i-1)+1;      % one step to the right
9      else
10          walk(i)=walk(i-1)-1;     % one step to the left
11      end
12  end
13  plot(walk,'k','linewidth',2);
14  hold on
```

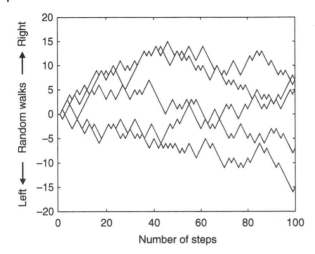

Figure 8.5 Stochastic simulations of the random walk for the drunkard, using Listing 8.6.

8.3 Monte Carlo Methods

Randomness plays an important role in many mathematical models, even when they are formulated deterministically. Monte Carlo methods are a broad class of computational algorithms that allow us to sample the solution space by generating random numbers [55, 19]. These methods of simulation, named after the Monte Carlo casino in Monaco, were first developed and used systematically during World War II to develop nuclear bombs [29]. While their range of application is vast, these methods are particularly useful for generating draws from a probability distribution, and simulating dynamical models with parameter variations. One pattern that is commonly used by many Monte Carlo methods is to model a system as probability density functions (PDFs), repeatedly sampling from the PDFs, and obtain the statistics related to the problem and model solutions. To illustrate the functionality of Monte Carlo methods, we consider a simple example of generating random numbers.

Example 8.5 Suppose X is a random variable that has an exponential distribution with parameter λ. We recall from Chapter 6 that the mean of this distribution is $1/\lambda$. The cumulative distribution function (CDF) for this random variable is $F(x) = 1 - e^{-\lambda x}$ for $x \geq 0$. To sample from the exponential distribution, we first sample a number r from the uniform distribution on the interval $[0, 1)$, and locate r on the y-axis of the CDF. We know from the inverse transform method that:

$$x = F^{-1}(r) = -\frac{\ln(1 - r)}{\lambda},$$

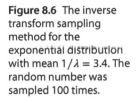

Figure 8.6 The inverse transform sampling method for the exponential distribution with mean $1/\lambda = 3.4$. The random number was sampled 100 times.

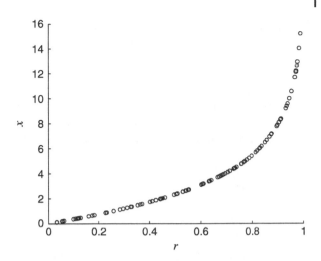

which gives the value of the random variable with the exponential distribution. Figure 8.6 shows the sampling of a random variable that has exponential distribution with mean $1/\lambda = 3.4$. To generate these samples, the Matlab© code in Listing 8.7 was used.

This random sampling can be used for stochastic simulations. In Example 8.3 we note that the waiting time for an event to occur is exponentially distributed. For example, if the recovery rate of infected individuals is γ, then the mean infectious period is $1/\gamma$. Thus, we can assume that the infectious period has an exponential distribution with mean $1/\gamma$. For the purpose of simulations, one needs to determine waiting times for all events that may be associated with different distributions. In the simulations for the SIR model in Example 8.3, we used an exponential distribution with mean $1/\sum_i a_i$ (i.e., equation (8.2)), where a_i is the rate of occurrence of event i. Thus, the random variable was generated for the waiting time or transition time during which an event occurs (which is defined as *step* in Listing 8.2) using `step=-log(Random1)/(sum(Rate))`.

Listing 8.7: Matlab© code for sampling a random variable with exponential distribution.

```
1  lambda=1/3.4;      % parameter of the exponential distribution
2  for i=1:100
3      r=rand;        % sample a number from uniform distribution
4      Rand(i)=r;     % vector of random numbers
5      x(i)=-log(1-r)/(lambda);  % vector of random variables
6  end
7  plot(Rand,x,'o')  % plot random variables
```

Exercises

1   Vaccination is an effective method of preventing infection for many diseases by stimulating the immune responses against the target pathogen. However, immune responses generally wane over time, and therefore the vaccine-induced immunity may become ineffective in disease prevention. A mathematical model representing the dynamics of vaccination and disease spread in the population can be expressed by:

$$\frac{dS}{dt} = (1-p)\Lambda - \beta SI - \mu S + \omega V,$$

$$\frac{dV}{dt} = p\Lambda - \mu V - \omega V + \gamma I,$$

$$\frac{dI}{dt} = \beta SI - \mu I - \gamma I,$$

where S, V, and I represent the population of susceptible, vaccinated, and infected individuals; Λ is the birth rate; p is the fraction of newborns who are vaccinated against the infection; β is the transmission rate of infection; ω is the rate of loss of vaccine-induced immunity or naturally acquired immunity through infection; γ is the rate of recovery from infection; and μ is the natural death rate. For this model, use the parameter values $\Lambda = 1000$, $\beta = 2 \times 10^{-5}$, $\mu = 0.02$, $\gamma = 73$, $\omega = 0.05$, $p \in [0, \ 1]$, with the initial conditions $S(0) = 45{,}000$, $V(0) = 4995$, and $I(0) = 5$ to:

a) develop a nonstandard finite-difference method to simulate the model for your choice of parameter p in the range $[0, \ 1]$;

b) develop a stochastic version of the model to simulate and compare the results with those obtained for the deterministic model using the same parameter values;

c) find the minimum value of p for which $I \rightarrow 0$ for large t using both deterministic and stochastic simulations;

d) determine the probability of no epidemic occurring in the absence of vaccination in a stochastic scenario.

2   The Brusselator system is an example of an autocatalytic, oscillating chemical reaction, in which a species acts to increase the rate at which it produces a reaction. The mechanisms for the Brusselator are given by:

$$A \xrightarrow{k_1} X,$$

$$B + X \xrightarrow{k_2} Y + C,$$

$$2X + Y \xrightarrow{k_3} 3X,$$

$$X \xrightarrow{k_4} D,$$

where X and Y are two intermediate species of chemicals with concentrations $[X]$ and $[Y]$ that come from an infinite supply in this reaction scheme. Molar concentrations of $[A]$ and $[B]$ are large and therefore are considered as constants in the reaction model. A mathematical model representing this reaction system for species X and Y is given by:

$$\frac{d[X]}{dt} = k_1[A] - k_2[B][X] + k_3[X]^2[Y] - k_4[X],$$

$$\frac{d[Y]}{dt} = k_2[B][X] - k_3[X]^2[Y],$$

where the k_i are the rates of the corresponding reactions. A dimensionless form of the system can be written as:

$$\frac{d[X]}{dt} = 1 - (1 + b)[X] + a[X]^2[Y],$$

$$\frac{d[Y]}{dt} = b[X] - a[X]^2[Y].$$

a) Analyze the model theoretically to find the conditions under which the system undergoes a Hopf bifurcation.
b) By changing the parameter values of a and b, explore the phase-plane dynamics of the model using both the Euler and nonstandard methods.
c) Show the existence of a stable limit cycle in simulations when the Hopf bifurcation occurs.

3 Food chains are important components of species interactions in ecosystems. A food-chain model can be developed as an extension of a predator–prey system by the following system of differential equations:

$$x' = rx(1 - x) - y\phi(x),$$
$$y' = y(\mu\phi(x) - d_1) - z\psi(y),$$
$$z' = z(\eta\psi(y) - d_2),$$

where x is the lowest trophic species (prey), y is the population size of the middle trophic level species (first predator), and z is the population size of the highest trophic level species (second predator). The parameter r is the intrinsic growth rate of species x; μ and η are the conversion rates of predation for species y and z, respectively; the constants d_1 and d_2 represent the death rates for each of these species; and $\phi(x)$ and $\psi(y)$ are the response functions of predators to their prey. The feeding relationships in this three-level food chain are such that the predations occur only upon the next lower adjacent level.

Consider the following response functions:

$$\phi(x) = \frac{x}{1+x}, \quad \psi(y) = \frac{y}{1+y}.$$

Develop a nonstandard finite-difference method to simulate the system. Also develop a Gillespie algorithm to perform stochastic simulations. For these methods, use $r = 1.1$, $\mu = 5$, $\eta = 0.006$, $d_1 = 0.2$, and $d_2 = 0.001$, with initial conditions $x(0) = 1$, $y(0) = 1.3$, and $z(0) = 3.3$.

4 Dengue is an important vector-borne disease that is endemic in many tropical regions [17, 54]. Worldwide, there are an estimated 2.5–3.6 billion individuals who are at risk of dengue infection, with annual estimates of 50–230 million new cases, 500,000 hospitalizations, and 25,000 fatal outcomes [54]. Dengue is caused by four antigenically distinct virus serotypes, designated as DENv1, DENv2, DENv3, and DENv4 [9]. These serotypes are transmitted from infectious individuals to susceptible individuals primarily through the bites of infectious mosquitoes, mainly *Aedes aegypti*.

Considering human and mosquito populations, develop a model that describes the transmission dynamics of dengue disease for a single virus serotype. A schematic diagram for infection dynamics is shown in the figure below. A susceptible mosquito can become infected from an infectious human [24]. There is an incubation period before the infected mosquito becomes infectious and transmits the disease to a susceptible human. Similarly, there is an incubation period for an infected individual before the start of the infectious period.

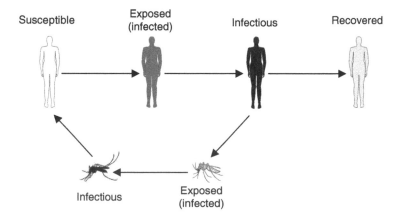

It has been shown that recovery from one serotype of dengue provides permanent immunity against the same serotype, but only partial immunity against other serotypes of dengue [40]. Extend your model to include two serotypes of dengue. For each model, develop a simulation algorithm to illustrate disease dynamics in the human population. Explore published studies about modelling dengue dynamics to extract plausible parameter values for your simulations. These parameters include transmission rates from human to mosquito and mosquito to human, average mosquito life-time, ratio of mosquito population to human population, average incubation periods for infected mosquitoes and humans, and the infectious period for infected humans. There is no recovery for infected mosquitoes.

9

Examples of Mathematical Modelling

In this chapter, we will present some examples of mathematical models with application to real-life phenomena. In these examples, we will use techniques learnt in previous chapters in analyzing the models.

9.1 Traffic Model

Suppose we have a one-way flow of traffic in one lane of a road (Figure 9.1). We consider the following assumptions [18]:

1) Cars are modelled by moving points.
2) The location of a car, denoted by x, is measured from some reference point along the x-axis (parallel to the road).
3) The traffic density, denoted by $\rho(x, t)$, depends on time and the location of the cars. It is defined as the number of cars per unit length of the road at location x and time t.

In this example, we are interested in determining the density $\rho(x, t)$ as a measure of traffic intensity. To do so, let us consider a section of the road $[x, x + \Delta x]$ at time t (Figure 9.1). Suppose that $\Delta N([x, x + \Delta x], t) = n_0$ is the number of cars in this section of the road. Thus, the traffic density is given by:

$$\rho(x, t) = \lim_{\Delta x \to 0} \frac{\Delta N([x, x + \Delta x], t)}{\Delta x} \quad \text{(if the limit exists).}$$

This equation can be used to approximate the number of cars as $\Delta N([x, x + \Delta x], t) \approx \rho(x, t)\Delta x$. Thus, for two distinct points a and b on the road, we have:

$$N(t) = \int_a^b \rho(x, t)dx. \tag{9.1}$$

Mathematical Modelling: A Graduate Textbook, First Edition. Seyed M. Moghadas and Majid Jaberi-Douraki.
© 2019 John Wiley & Sons, Inc. Published 2019 by John Wiley & Sons, Inc.
Companion Website: www.wiley.com/go/Moghadas/Mathematicalmodelling

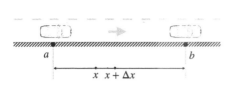

Figure 9.1 Representation of traffic in a single lane.

We assume that $\rho(x, t)$ is a differentiable function. Using the balance law in modelling, we have:

$$\frac{dN}{dt} = \{\text{influx}\} - \{\text{outflux}\}$$

$$= \{\text{number of cars entering at point } a\}$$

$$- \{\text{number of cars leaving at point } b\}.$$

Suppose that the traffic flux at any location x at time t is $q(x, t)$, which measures the number of cars passing through the point x per unit time. Thus,

$$\frac{dN}{dt} = q(a, t) - q(b, t).$$

We also make the assumption that $q(x, t)$ is differentiable and its partial derivatives with respect to x and t are continuous. From the fundamental theorem of calculus, it follows that:

$$\frac{dN}{dt} = -\int_a^b \frac{\partial}{\partial x}(q(x, t))dx. \tag{9.2}$$

Also, from (9.1), we get:

$$\frac{dN}{dt} = \int_a^b \frac{\partial}{\partial t}(\rho(x, t))dx. \tag{9.3}$$

Comparing (9.2) and (9.3) yields:

$$\int_a^b \frac{\partial}{\partial t}(\rho(x, t))dx + \int_a^b \frac{\partial}{\partial x}(q(x, t))dx = 0,$$

or

$$\int_a^b \left(\frac{\partial}{\partial t}(\rho(x, t)) + \frac{\partial}{\partial x}q(x, t) \right) dx = 0.$$

Since a and b are arbitrary points on the road, we have the following partial differential equation:

$$\frac{\partial}{\partial t}\rho(x, t) + \frac{\partial}{\partial x}q(x, t) = 0. \tag{9.4}$$

Our aim is to solve (9.4) for $\rho(x, t)$ with some initial condition $\rho(x, 0) = \rho_0(x)$. To this end, we first express $q(x, t)$ in terms of $\rho(x, t)$. Suppose that a car is moving with velocity v. Thus, for a short time interval $[t, t + \Delta t]$, the car moves a distance of approximately $\Delta x \approx v \Delta t$. The number of cars in $[x, x + \Delta x]$ is therefore given by:

$$\Delta N \approx \rho(x, t) v \Delta t,$$

which gives $\frac{\Delta N}{\Delta t} = \rho v$. Taking the limit as $\Delta t \to 0$, we get:

$$\lim_{\Delta t \to 0} \frac{\Delta N}{\Delta t} = q(x, t) = \rho(x, t) v.$$

Summarizing the above, we get the equation:

$$\frac{\partial \rho}{\partial t} + \frac{\partial (\rho v)}{\partial x} = 0. \tag{9.5}$$

In order to find the solution to (9.5), we need a model for the velocity v. Let $v(\rho) = f(\rho, \Lambda)$, where Λ is a set of constant parameters. If the density is small, then cars can go very fast (i.e., $v = v_{max}$). If the density is at its maximum ρ_{max}, then cars cannot move (there is a traffic jam) and we have $v = 0$. This process may be modelled by the following equation:

$$v(\rho) = v_{max} \left(1 - \frac{\rho}{\rho_{max}} \right).$$

Substituting v into (9.5) provides a partial differential equation for ρ. To solve this equation, we first introduce the following scaling variables:

$$u(x, t) = \frac{\rho(x, t)}{\rho_{max}}, \quad y = \frac{x}{L}, \quad \tau = \frac{t}{\kappa},$$

where L and κ are scaling factors for distance and time. Using the chain rule, we get:

$$\frac{\partial \rho}{\partial x} = \frac{\partial \rho}{\partial y} \cdot \frac{\partial y}{\partial x} = \frac{1}{L} \frac{\partial \rho}{\partial y},$$

$$\frac{\partial \rho}{\partial t} = \frac{\partial \rho}{\partial \tau} \cdot \frac{\partial \tau}{\partial t} = \frac{1}{\kappa} \frac{\partial \rho}{\partial \tau}.$$

From (9.5) we have:

$$\frac{1}{\kappa} \frac{\partial \rho}{\partial \tau} + \frac{v_{max}}{L} \frac{\partial}{\partial y} ((1 - u)\rho) = 0.$$

Dividing this equation by ρ_{max} and multiplying by κ, we get:

$$\frac{\partial u}{\partial \tau} + \left(\frac{v_{max} \kappa}{L} \right) \frac{\partial}{\partial y} ((1 - u)u) = 0.$$

Suppose $\kappa = L/v_{max}$. Thus, we get the dimensionless equation:

$$\frac{\partial u}{\partial \tau} + \frac{\partial}{\partial y} ((1 - u)u) = 0. \tag{9.6}$$

If t and x are assumed to be in units of κ and L, respectively, then we can rewrite (9.6) in the form:

$$\frac{\partial u}{\partial t} + \frac{\partial}{\partial x}((1-u)u) = 0. \tag{9.7}$$

This equation can be solved using the method of characteristics, where we assume $g(u) = (1-u)u$, and $\frac{dx}{dt} = g'(u)$. This implies that $\frac{du}{dt} = 0$, and therefore the solution is constant along the characteristics. Assuming that $g'(u)$ exists and is continuous, we have $g'(u) = 1 - 2u$, and therefore,

$$\begin{aligned}
\frac{du}{dt} &= \frac{\partial u}{\partial t} \cdot \frac{dt}{dt} + \frac{\partial u}{\partial x} \cdot \frac{dx}{dt} \\
&= \frac{\partial u}{\partial t} + (1-2u)\frac{\partial u}{\partial x} \\
&= \frac{\partial u}{\partial t} + \frac{\partial u}{\partial x}(g(u)) = 0.
\end{aligned}$$

We therefore find that $u = \phi(k)$, where $\phi(k)$ is a constant and k labels the different characteristics. Also, along the characteristics, we have:

$$\frac{dx}{dt} = 1 - 2\phi(k),$$

which gives a solution $x(t) = (1 - 2\phi(k))t + k$. Thus, $k = x(t) - (1 - 2\phi(k))t = x - (1 - 2u)t$. This gives a solution that defines u implicitly by $u(x, t) = \phi(x - (1 - 2u)t)$. Using the initial condition at $t = 0$, we have $u(x, 0) = \rho_0(x)/\rho_{max}$, and hence $\phi(x) = \rho_0(x)/\rho_{max}$. For example, if $\rho_0(x) = \rho_{max}x$, then $\phi(x) = x$ and the solution for $u(x, t)$ is $u(t, x) = x - (1 - 2u(t, x))t$. Thus, we get:

$$u(x, t) = \frac{x - t}{1 - 2t}, \quad x \in \mathbb{R}, \quad 0 \le t < \frac{1}{2}.$$

9.2 Michaelis–Menten Kinetics

Enzymatic reactions are an important mechanism in biological systems [38]. Here, we consider a classic model of enzyme reaction and discuss its kinetics, where the reaction rate can be obtained at various concentrations of the enzyme and substrate. Formation of an enzyme–substrate complex is the key component of reaction systems. In 1913, Leonor Michaelis and Maud Menten suggested that the binding of the substrate and the enzyme is reversible, and derived a model to describe the kinetics of a single-substrate enzymatic reaction [36].

Consider a single-substrate reaction, where the free enzyme E binds to the substrate S to form a complex ES before catalyzing the reaction, and then dissociating from the product P. This chemical process is described by:

$$E + S \underset{k_{-1}}{\overset{k_1}{\rightleftharpoons}} E \cdot S \xrightarrow{k_2} E + P, \tag{9.8}$$

where k_1 is the rate of complex formation, k_{-1} is the rate of complex breakdown, and k_2 is the rate of dissociation. The molar concentration for each component of the reaction system is denoted by $[\cdot]$. Throughout the reaction, the total concentration of the enzyme will be the sum of the concentration of total free enzyme $[E]$ and the concentration of total enzyme bound with the substrate $[ES]$. Initially, the substrate concentration is more than that of the enzyme concentration. As reaction proceeds, the concentration of the complex increases with time and reaches a steady state. To derive the Michaelis–Menten model, we assume that the concentration of the enzyme–substrate complex remains the same over time once the system reaches the steady state. This implies that:

$$\frac{d[ES]}{dt} = 0.$$

The free enzyme $[E]$ at any point of time in the reaction is obtained from the difference between the concentration of the enzyme–substrate complex $[ES]$ and the total enzyme concentration $[E_{total}]$,

$$[E] = [E_{total}] - [ES].$$

From (9.8) we have the following rates:

$$\text{rate of } [ES] \text{ formation} = k_1([E_{total}] - [ES])[S],$$
$$\text{rate of } [ES] \text{ breakdown} = k_{-1}[ES] + k_2[ES].$$

At the steady state of the system, the rate of $[ES]$ formation is the same as its rate of breakdown, and therefore we get:

$$[ES] = \frac{[E_{total}][S]}{k_m + [S]}, \tag{9.9}$$

where

$$k_m = \frac{k_{-1} + k_2}{k_1}.$$

From (9.8), we also obtain the rate of change in the product over time as:

$$\frac{d[P]}{dt} = k_2[ES]. \tag{9.10}$$

Substituting (9.9) into (9.10) gives the classical form of the Michaelis–Menten equation:

$$\frac{d[P]}{dt} = \frac{k_2[E_{total}][S]}{k_m + [S]}. \tag{9.11}$$

The rate $k_2[E_{total}]$ is often written as V_{max}, as it represents the rate of product formation provided that all enzyme is bound to substrate. From (9.11), it follows that at a very high concentration of the substrate, $\frac{d[P]}{dt} \approx V_m$ and the product increases with time independent of the substrate. Furthermore, if the substrate concentration is very low, then $\frac{d[P]}{dt} \approx V_m[S]/k_m$.

Considering the balance law, we can also present a deterministic system of differential equations for the reaction displayed in (9.8):

$$\frac{d[E]}{dt} = -k_1[E][S] + (k_{-1} + k_2)[ES],$$

$$\frac{d[S]}{dt} = -k_1[E][S] + k_{-1}[ES],$$

$$\frac{d[ES]}{dt} = k_1[E][S] - (k_{-1} + k_2)[ES],$$

$$\frac{d[P]}{dt} = k_2[ES].$$

The critical points of this model are $(0, [S^*], 0, [P^*])$, $([E^*], 0, 0, [P^*])$ and $(0, 0, 0, [P^*])$. This means that at the critical points, there is no complex and the concentration of enzyme or substrate (or both) has been exhausted. We leave the analysis of these critical points as an exercise.

9.3 The Brusselator System

The Brusselator is a system of differential equations which describe an autocatalytic oscillating chemical process in which a species acts to increase the rate at which it produces a reaction [14]. The reaction mechanisms for the Brusselator system are described by:

$$\begin{aligned} A &\to X, \\ B + X &\to Y + C, \\ 2X + Y &\to 3X, \\ X &\to D \end{aligned} \tag{9.12}$$

where A and B are input chemical species, C and D are reaction products, and X and Y are two autocatalytic species. These mechanisms can be mathematically expressed by the following system of differential equations [52]:

$$\begin{aligned} \frac{dX}{dt} &= B + X^2Y - (1 + A)X \equiv F(X, Y), \\ \frac{dY}{dt} &= AX - X^2Y \equiv G(X, Y), \end{aligned} \tag{9.13}$$

where $A, B > 0$. This system has a unique critical point $E_c = (B, A/B)$. The Jacobian of the system at E_c is:

$$J_{E_c} = \begin{bmatrix} A - 1 & B^2 \\ -A & -B^2 \end{bmatrix},$$

with the characteristic equation:

$$\lambda^2 - (A - 1 - B^2)\lambda + B^2 = 0.$$

The eigenvalues of the system are:

$$\lambda_{\pm} = \frac{(A - 1 - B^2) \pm \sqrt{(A - 1 - B^2)^2 - 4B^2}}{2}.$$

If $A - 1 - B^2 < 0$, then the eigenvalues $\lambda_{\pm}$ have negative real parts. Thus, E_c is locally asymptotically stable. If $A - 1 - B^2 > 0$, then the eigenvalues have positive real parts, and hence E_c is unstable.

Now suppose $A - 1 - B^2 = 0$. Thus, the eigenvalues are purely imaginary, $\lambda_{\pm} = \pm B$ i. Since A, $B > 0$, this case can occur only if $A > 1$. For any fixed $A > 1$, let $\alpha(B) = -(A - 1 - B^2)$ and $(A - 1 - B^2)^2 - 4B^2 \neq 0$. Taking the derivative of α with respect to B and calculating at $\sqrt{A - 1}$ gives $\alpha'(B) = 2\sqrt{A - 1} > 0$. Some algebraic calculations show that at E_c, we have:

$$
\begin{aligned}
a = & \frac{1}{16}\left(F_{XXX} + F_{XYY} + G_{XXY} + G_{YYY}\right)\Big|_{(E_c, B^2 = A-1)} \\
& + \frac{1}{16\omega}\left(F_{XY}(F_{XX} + F_{YY}) - G_{XY}(G_{XX} + G_{YY})\right. \\
& \left. -F_{XX}G_{XX} + F_{YY}G_{YY}\right)\Big|_{(E_c, B^2 = A-1)} \\
= & -\frac{1}{8}\left(1 - \frac{A^2}{B^3}\right) = -\frac{1}{8}\left(\frac{B^3 - (B^2 + 1)^2}{B^3}\right) > 0.
\end{aligned}
$$

This shows that the conditions for the occurrence of a Hopf bifurcation are satisfied (see Chapter 4). Since $a\alpha'(B) > 0$, a unique curve of stable periodic solutions bifurcates at the bifurcation parameter. Figure 9.2 shows the changes in the system behavior when the stable E_c becomes unstable, and a stable limit cycle appears for different values of A and B.

9.4 Generalized Richards Model

In 1959, Richards [41] extended the logistic model given by (2.5) in Chapter 2 to include a parameter that measures the extent of deviation from the S-shaped dynamics of the classical logistic growth model. The Richards model is expressed by the following differential equation:

$$N' = rN\left[1 - \left(\frac{N}{K}\right)^a\right], \tag{9.14}$$

where $a > 0$, and r and K are the parameters of the logistic equation. The explicit solution of the Richards model is:

$$N(t) = \frac{K}{[1 + e^{-ra(t-\tau)}]^{\frac{1}{a}}}, \tag{9.15}$$

(a)

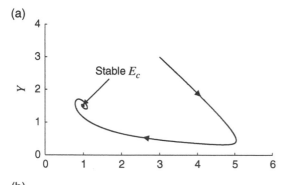

(b)

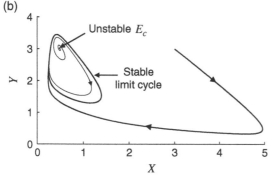

Figure 9.2 Simulations for the Brusselator system with (a): $A = 1.5$ and $B = 1$ (for which the solutions approach the stable critical point E_c); and (b) $A = 1.5$ and $B = 0.5$ (for which the solutions approach the stable limit cycle).

where τ is the time for the change in the growth rate. This model describes the population dynamics in two phases of fast and slow growth with a transition point (also referred to as the turning point or inflection point). In the slow phase of growth (after the turning point), the population approaches its carrying capacity K. At the turning point, one can obtain the population size as $N = K/(1 + a)^{1/a}$. Here, we generalize the Richards model to the following form:

$$N' = rN^p \left[1 - \left(\frac{N}{K} \right)^a \right],$$
(9.16)

where p is a parameter in the range $(0, 1]$. If $p = a = 1$, then we obtain the classical logistic equation. For (9.16), one can see that if the population size is very small compared to the carrying capacity K (i.e., $N/K \ll 1$), then the population growth can be estimated by $N' = rN^p$. When $p = 1$, this growth is exponential and $N(t) = N(0)e^{rt}$. However, when $p < 1$, the population has polynomial growth (see Exercise 1 below). For the generalized Richards model, by taking an implicit derivative of (9.16) and solving $N'' = 0$ we obtain:

$$N(\text{turning point}) = K \left(\frac{p}{p + a} \right)^{\frac{1}{a}}.$$
(9.17)

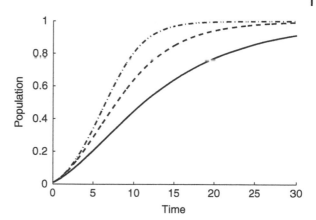

Figure 9.3 The solution curves of the Richards model with $r = 0.2$ and $K = 1$ for different values of a: solid curve ($a = 0.5$); dashed curve ($a = 1$); and dot-dashed curve ($a = 2$).

Using (9.17), one can estimate the turning point of the growth curve. To do so, we first consider the change of variable $x = (N/K)^a$, and transform (9.16) into the form $x' = raK^{p-1}x^\delta(1 - x)$, where $\delta = 1 + (p - 1)/a$. Since $x < 1$, one can use geometric series to solve this equation using the method of separation and integrating from time $t = 0$ to $t = t_1$. Thus, we get:

$$\sum_{n=0}^{\infty} \left(\frac{N(t_1)^{na+1-p} - N(0)^{na+1-p}}{(na + 1 - p)K^{na+1-p}} \right) = rK^{p-1}t_1. \tag{9.18}$$

Substituting (9.17) into (9.18), the turning point t_{turn} of the growth curve can be calculated as:

$$t_{\text{turn}} = \frac{1}{rK^{p-1}} \sum_{n=0}^{\infty} \frac{\left(\frac{p}{p+a} \right)^{n+\frac{1-p}{a}} - \left(\frac{N(0)}{K} \right)^{na+1-p}}{na + 1 - p}. \tag{9.19}$$

We note that in the logistic equation (i.e., $a = p = 1$), the turning point occurs when the population reaches half of its carrying capacity (i.e., $N = K/2$). In the generalized Richards model, the turning point depends on the values of p and a. Figure 9.3 shows three different growth curves for the generalized Richards model simulated using the Euler method. To simulate (9.16), we used the Matlab© code given in Listing 9.1.

Listing 9.1: Matlab© code for simulating the generalized Richards model.

```
1  %% Parameters of the system (can vary)
2  K=1;
3  r=0.2;
4  a=0.5;
5  p=0.5;
6  %% Creating a vector of time for discritization
7  T0=0;          % T0 is the beginning of time interval
8  Tend=30;       % Tend is the end of time interval
9  h=0.01;        % h is a fixed step-size for discretiziation
```

```
10  t=0:h:30;          % creating the time vector
11  Lt=length(t);      % Lt is the length of time vector
12  %% Creating vectors of system variable
13  N=zeros(Lt,1);     % creates a vector of Lt by 1 zeros for N
14  %% Initial conditions (can vary)
15  N(1)=0.01;
16  %% Euler method
17  for j=2:Lt
18      N(j)=N(j-1)+h*r*N(j-1)^p-h*r*N(j-1)^p*(N(j-1)/K)^a;
19  end
20  %% Plotting solution N as a function of time
21  plot(t,N,'k')
```

9.5 Spruce Budworm Model

Most species of spruce budworms are pests that destroy coniferous forests. *Choristoneura fumiferana* is the most serious defoliator of spruce and balsam fir trees in the boreal forests of the United States and Canada [28]. The impact of these species has been felt in periodic outbreaks, most notably in the twentieth century, between 1910–1920, 1940–1950, and 1970–1980 in eastern Canada. The ecological dynamics of these outbreaks has been studied using mathematical models [23, 42] that include the interactions between the spruce budworm population with its host, the coniferous trees, and its primary predator, birds.

A simple model describing the budworm population size over time can be built using the logistic growth model [26]. Denoting the budworm population size by B, the model can be expressed by the following differential equation:

$$\frac{dB}{dt} = rB\left(1 - \frac{B}{K}\right) - \frac{aB^2}{b^2 + B^2}, \tag{9.20}$$

where r is the growth rate of budworms, K is the carrying capacity limiting the growth of the budworm population in the presence of limited resources, and

$$\frac{aB^2}{b^2 + B^2},$$

represents the rate of predation of budworms by birds. To analyze this model, we first introduce new scaling variables to nondimensionalize the system. Let:

$$P = \frac{B}{b}, \quad s = \frac{at}{b}.$$

Thus, the system reduces to:

$$\frac{dP}{ds} = \frac{rb}{a}P\left(1 - \frac{bP}{K}\right) - \frac{P^2}{1 + P^2}. \tag{9.21}$$

Figure 9.4 Representation of the critical points for model (9.21). Dashed lines correspond to different values of rb/a.

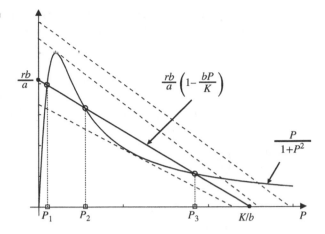

A critical point of the system is at $P = 0$. Taking the Jacobian gives:

$$J = \frac{rb}{a} - \frac{2rb^2 P}{aK} - \frac{2P}{(1+P^2)^2} = \frac{rb}{a}\left(1 - \frac{2bP}{K}\right) - \frac{2P}{(1+P^2)^2}.$$

At $P = 0$, the Jacobian is $J_0 = rb/a > 0$, and therefore the critical point is unstable. Other critical points of the reduced model when $P \neq 0$ can be found by solving the equation:

$$\frac{rb}{a}\left(1 - \frac{bP}{K}\right) = \frac{P}{1+P^2}.$$

Depending on the parameter values, this equation can have one, two, or three solutions, as depicted in Figure 9.4. For example, for a sufficiently high value of rb/a the system has only one equilibrium if the slope of the line in Figure 9.4 is very small or very large (dashed lines in Figure 9.4). Thus, choosing r as the bifurcation parameter, one can explore the number of critical points and the type of bifurcation the system undergoes. The stability of these critical points can also be investigated using numerical simulations, and we leave these tasks as an exercise.

9.6 FitzHugh–Nagumo Model

In 1948–1952, Hodgkin and Huxley conducted a series of patch clamp experiments on the giant squid axon to develop a mathematical model for describing the variation in concentration of sodium ions and potassium ions across the cell membrane of a nerve cell, the so-called neuron. A neuron is an electrically excitable cell that processes and transmits information through electrical and chemical signals. A simpler form of the Hodgkin–Huxley model was formulated later by FitzHugh in 1961 [13] and Nagumo in 1962 [35]. The FitzHugh–Nagumo model governs the dynamics of electrical activity in a neuron, known as an excitable system. A neuron can be stimulated and subsequently *excited* by a flow of electric charge. When a neuron is excited, physiological processes in the cell can recover the neuron from the excitation. Denoting by x and y respectively the excitation and recovery as functions of time, a form of the FitzHugh–Nagumo model is expressed by the following system of differential equations:

$$\frac{dx}{dt} = c\left(y + x - \frac{x^3}{3}\right) + I,$$
$$\frac{dy}{dt} = \frac{a - by - x}{c},$$

(9.22)

where I is a constant input to the system, and other parameters are positive satisfying the following assumptions:

$$1 - \frac{2b}{3} < a < 1, \quad 0 < b < 1, \quad b < c^2.$$

If $I = 0$, then the critical points of the model are the solutions of the equation:

$$\frac{bx^3}{3} + (1 - b)x - a = 0.$$

The left-hand side of this equation is an increasing polynomial of degree 3, and therefore has only one positive solution. Let x_* be the solution of this cubic equation. From the conditions on the parameters, it can be shown that $x_* < 1 - 2b/3$, and therefore $y^* = (a - x_*)/b > 0$. The Jacobian of (9.22) at (x_*, y_*) is:

$$J_{(x_*, y_*)} = \begin{bmatrix} c(1 - x_*^2) & c \\ -\frac{1}{c} & -\frac{b}{c} \end{bmatrix}.$$

Since $1 - b + bx_*^2 > 0$, the stability of the critical point depends on the sign of :

$$D = c(1 - x_*^2) - \frac{b}{c}.$$

If $D < 0$, then the eigenvalues of $J_{(x_*, y_*)}$ have negative real parts and the critical point is a stable node (or sink). If $D > 0$, then the eigenvalues have positive real parts and the critical point is an unstable node (or source). If $D = 0$, then the eigenvalues are purely imaginary (complex numbers). In this case, $x_*^2 = 1 - b/c^2$. Considering $\alpha(D)$ as the real part of the eigenvalues, we now test the conditions for a Hopf bifurcation. Clearly,

$$d = \frac{d\alpha(D)}{dD}\bigg|_{D=0} = \frac{1}{2} > 0.$$

To establish the type of Hopf bifurcation, we use parameter values $a = 0.7$ and $b = 0.8$. Thus, $x_* = 1.1994$ and $y_* = -0.6249$. It follows from the expression for D that the Hopf bifurcation occurs for $c_* = 0.4228$. It can be seen that the Hopf bifurcation is supercritical for $c > c_*$. Figure 9.5 shows the stable limit cycle resulting from the Hopf bifurcation using a nonstandard finite-difference method. To simulate (9.22), we used the Matlab© code given in Listing 9.2. We leave the model analysis for the case of nonzero input (i.e., $I \neq 0$) as an exercise.

Listing 9.2: Matlab© code for simulating the FitzHugh–Nagumo model.

```
1  %% Parameters of the system
2  a=0.7;
3  b=0.8;
4  c=3;
5  %% Creating a vector of time for discritization
6  T0=0;              % T0 is the beginning of time interval
7  Tend=200;          % Tend is the end of time interval
8  h=0.001;           % h is a fixed step-size for discretiziation
9  t=0:h:200;         % creating the time vector
10 Lt=length(t);      % Lt is the length of time vector
11 %% Creating vectors of system variables
12 x=zeros(Lt,1);     % creates a vector of Lt by 1 zeros for x
13 y=zeros(Lt,1);     % creates a vector of Lt by 1 zeros for y
14
15 %% Initial conditions (can vary)
16 x(1)=0;    % x0 is the initial value of x at time T0
17 y(1)=0;    % y0 is the initial value of y at time T0
18 %% Non-standard finite-difference method
19 for j=2:Lt
20     x(j)=(x(j-1)+c*h*(x(j-1)+y(j-1)))/(1+h*c*(x(j-1))^2/3);
21     y(j)=(y(j-1)+h*(a-x(j-1))/c)/(1+h*b*y(j-1)/c);
22 end
23 %% Plotting phase plane (limit cycle)
24 plot(x,y,'r');
```

9.7 Decay Model

Suppose $N(t)$ is the quantity of radioactive nuclei at time t (Example 2.1 in Chapter 2). Given the decay rate of K nuclei per unit time, we are interested in determining the probability that there are n surviving nuclei (i.e., they are still in the compartment) at time t. Let $p_n(t)$ represent this probability. For the decay rate K, the decay probability for one nucleus is $K\Delta t + \mathcal{O}(\Delta t^2)$ during the transition time Δt. The probability $p_n(t)$ has the following components:

1) If one reaction has occurred during the transition time Δt,

$$p_n^{(1)}(t + \Delta t) = p_{n+1}(t)(n + 1)K\Delta t + \mathcal{O}(\Delta t^2). \tag{9.23}$$

2) If no reaction has occurred during the transition time Δt,

$$p_n^{(2)}(t + \Delta t) = p_n(t)(1 - nK\Delta t) + \mathcal{O}(\Delta t^2). \tag{9.24}$$

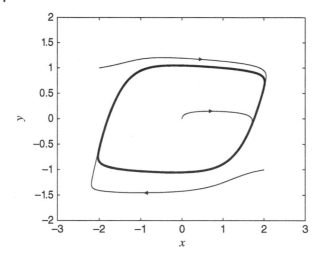

Figure 9.5 Simulations of the FitzHugh–Nagumo model with $a = 0.7, b = 0.8, c = 3$, and $I = 0$. The solid curve shows the stable limit cycle generated by the Hopf bifurcation.

3) If more than one reaction has occurred during the transition time Δt,

$$p_n^{(3)}(t + \Delta t) = \mathcal{O}(\Delta t^2). \tag{9.25}$$

Thus, we have:

$$p_n(t + \Delta t) = p_{n+1}(t)(n + 1)K\Delta t + p_n(t)(1 - nK\Delta t) + \mathcal{O}(\Delta t^2). \tag{9.26}$$

Rearranging this equation and taking the limit as $\Delta t \to 0$, we obtain the master equation for the decay model:

$$\frac{dp_n}{dt} = p_{n+1}(t)(n + 1)K - p_n(t)nK. \tag{9.27}$$

If the initial number of nuclei is N_0, then equation (9.27) is valid for $n = 0, 1, 2, \ldots, N_0 - 1$. However, for $n = N_0$, we have:

$$\frac{dp_{N_0}}{dt} = -N_0 K p_{N_0}(t), \quad p_{N_0}(0) = 1.$$

The equation for p_{N_0} can be solved to obtain the solution $p_{N_0}(t) = e^{-N_0 K t}$. This implies that when $t \to \infty, p_{N_0} \to 0$, and therefore the chance of having the initial quantity of nuclei in the compartment decreases over time. Substituting this solution back into (9.27), we get:

$$\frac{dp_{N_0-1}}{dt} = N_0 K e^{-N_0 K t} - p_{N_0-1}(t)(N_0 - 1)K.$$

This equation has solution:

$$p_{N_0-1}(t) = N_0 e^{-(N_0-1)Kt}(1 - e^{-Kt}).$$

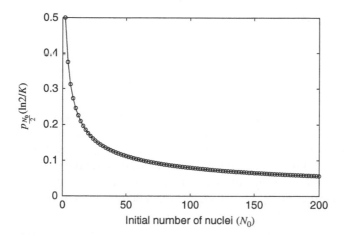

Figure 9.6 Probability that 50% of the nuclei are still in the compartment at time $t = \ln 2/K$ as a function of the initial number of nuclei.

By an induction argument, we find that the solution to (9.27) is:

$$p_n(t) = \binom{N_0}{n} e^{-nKt}(1 - e^{-Kt})^{N_0-n}, \quad n = 0, 1, 2, \dots, N_0.$$

Suppose $t = \ln 2/K$, which we obtained as the half-life of the nuclei in Example 2.1. In a deterministic context, the amount of material left in the compartment at time $\ln 2/K$ is $N_0/2$. However, in the stochastic model, the probability that $N_0/2$ nuclei are still in the compartment at time $t = \ln 2/K$ depends on N_0, and is given by:

$$p_{\frac{N_0}{2}}(\ln 2/K) = \binom{N_0}{\frac{N_0}{2}} e^{-N_0 \ln 2/2}(1 - e^{-\ln 2})^{N_0/2} = \binom{N_0}{\frac{N_0}{2}} \left(\frac{1}{4}\right)^{\frac{N_0}{2}}.$$

Figure 9.6 shows the curve of probability $p_{\frac{N_0}{2}}(\ln 2/K)$ for N_0 in the range [2,200], which represents a decreasing function of N_0. To simulate this probability, the Matlab© code in Listing 9.3 was used.

Listing 9.3: Matlab© code for simulating the decay model.

```
1  for i=1:100
2      n(i)=2*i;      % initial number of nuclei
3      N0=n(i);
4      prob(i)=nchoosek(N0,N0/2)*(1/4)^(N0/2); % probability
5  end
6  plot(n,prob,'.')   % plot probability as a function of N0
```

9.8 The Gambler's Ruin

Consider a gambler who starts with an initial fortune and plays a game to win more money. On each gamble, there are two possibilities: *win* which adds $1 to the gambler's fortune; and *lose* which reduces the gambler's fortune by $1. Suppose the total sum of money in this game is N. The game ends when the gambler's fortune is either $0 (and the gambler can no longer play the game) or $N (and the gambler has won all the money in the game). The term *ruin* refers to the situation in which the gambler's fortune is $0. In this game, it is assumed that the win and loss probabilities in each gamble are constants p and $q = 1 - p$ respectively, and independent of the outcomes of past gambles (Figure 9.7).

In the gambler's ruin problem [3], we are interested in determining the probability that the gambler wins all the money. Let P_k denote the probability that the gambler wins the entire N, having started with k. Obviously, $P_0 = 0$ (as the gambler has no fortune to play the game), and $P_N = 1$ (as the gambler has already gained the full N). If the gambler starts with $1, and the probability of winning all the money is P_2, then $P_1 = pP_2 + qP_0 = pP_2$. If the gambler starts with $N - 1$, then $P_{N-1} = pP_N + qP_{N-2}$. Now suppose the gambler starts with k, where $1 < k < N$. Then the gambler's fortune increases to $k + 1$ with probability p and decreases to $k - 1$ with probability q. Thus, $P_k = pP_{k+1} + qP_{k-1}$. This equation results from the following conditional probability:

$$P(E_k) = P(E_k| \text{ win})P(\text{win}) + P(E_k| \text{ lose})P(\text{lose}),$$

where E_k is the event that the gambler will go on to win the game. Here, we have denoted $P(E_k| \text{ win})$ by P_{k+1} and $P(E_k| \text{ lose})$ by P_{k-1}. Using the relation $p + q = 1$, we obtain $(p + q)P_k = pP_{k+1} + qP_{k-1}$, which can be written as:

$$P_{k+1} - P_k = \frac{q}{p}(P_k - P_{k-1}), \quad 1 < k < N - 1.$$

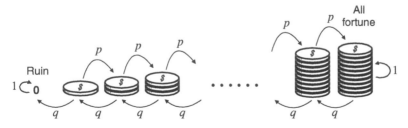

Figure 9.7 Representation of the gambler's ruin model.

Suppose $p \neq q$. Using a recursive calculation, we find that:

$$P_2 - P_1 = \frac{q}{p}(P_1 - P_0) = \frac{q}{p}P_1,$$

$$P_3 - P_2 = \frac{q}{p}(P_2 - P_1) = \left(\frac{q}{p}\right)^2 P_1,$$

$$\vdots$$

$$P_N - P_{N-1} = \frac{q}{p}(P_{N-1} - P_{N-2}) = \left(\frac{q}{p}\right)^{N-1} P_1.$$

Solving for P_k, $k \geq 1$, gives:

$$P_k = \left[1 + \frac{q}{p} + \left(\frac{q}{p}\right)^2 + \cdots + \left(\frac{q}{p}\right)^{k-1}\right] P_1 = \left[\frac{1 - \left(\frac{q}{p}\right)^k}{1 - \frac{q}{p}}\right] P_1.$$

In the case where $p = q = 1/2$, we have:

$$1 + \frac{q}{p} + \left(\frac{q}{p}\right)^2 + \cdots + \left(\frac{q}{p}\right)^{k-1} = k,$$

and therefore $P_k = kP_1$. Since $P_N = 1$, for $k = N$, we have:

$$P_1 = \begin{cases} \frac{1 - \frac{q}{p}}{1 - \left(\frac{q}{p}\right)^N}, & \text{if } p \neq q, \\ \frac{1}{N}, & \text{if } p = q = \frac{1}{2}. \end{cases}$$

Summarizing the above, the probability of winning \$N, having started with \$k, is:

$$P_k = \begin{cases} \frac{1 - \left(\frac{q}{p}\right)^k}{1 - \left(\frac{q}{p}\right)^N}, & \text{if } p \neq q, \\ \frac{k}{N}, & \text{if } p = q = \frac{1}{2}. \end{cases}$$

From the above formulation, one can see that if the gambler starts with \$k, then there is positive probability of becoming infinitely rich if $p > 1/2$. This is given by:

$$\lim_{N \to \infty} P_k = 1 - \left(\frac{q}{p}\right)^k.$$

On the other hand, if $p \leq 1/2$, then $\lim_{N \to \infty} P_k = 0$, and the gambler will be ruined.

Exercises

1   Suppose $r > 0$ and $0 < p < 1$. Show that the model $N'(t) = rN^p(t)$ represents the growth of a population in a polynomial form, and determine the degree of this polynomial. The initial population size at time $t = 0$ is $N_0 = N(0)$.

2   Consider Michaelis–Menten kinetics and introduce new variables to convert the model into a dimensionless system. Then analyze its critical points.

3   For the following deterministic SI epidemic model, develop a stochastic model:
$$S' = -\beta SI + \delta I,$$
$$I' = \beta SI - \delta I.$$

Clearly define all the variables and parameters used in the stochastic model. Develop a stochastic algorithm for simulating the model, and perform experiments with a reasonable choice of parameters.

4   Formulate a mathematical model using a system of differential equations that describe the replication of HIV viruses as shown in the figure below. The system should include uninfected cells (x), infected cells (y), and HIV viruses (v). Uninfected cells are produced at a constant rate b (independent of any other variables) and die at the rate d. These cells become infected through interactions with viruses at the rate β. Infected cells produce viruses at the rate p and die at the rate a. Viruses die at the rate c. For this model, determine the critical points and analyze their stability.

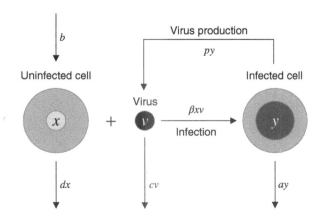

5   Consider the cascade of compartments shown in the following figure for
    the spread of an epidemic in the population. In this cascade, S, I, and V
    represent the populations of susceptible, infected, and vaccinated indi-
    viduals, respectively. Suppose B is the constant recruitment rate of the
    population; β is the transmission rate of infection between infected and
    susceptible individuals; μ is the natural death rate; ξ is the rate at which
    susceptible individuals are vaccinated; γ is the recovery rate from infec-
    tion; and δ is the rate of loss of immunity after recovery or vaccination.
    a) Model this cascade mathematically, and find the critical point at which
       there is no infection.
    b) Find the condition under which an epidemic is expected to be eradi-
       cated. (*Hint*: Determine the stability of the critical point at which there
       is no infection.)

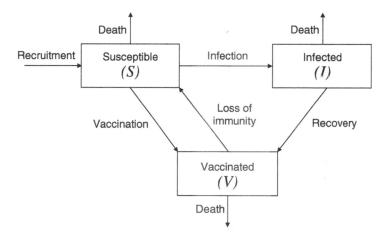

6   What is the one-step transition probability matrix in the gambler's ruin
    problem with sample space $S = \{0, 1, \dots, a\}$?

7   Suppose a gambler starts with $5 and aims to win $20 by flipping coins.
    Each flip will win $1 if the gambler gets heads; otherwise the gambler loses
    $1. What is the probability of the gambler winning $20? We assume that
    a fair coin is used. What is the probability of winning if the game involves
    using a die and the gambler wins $1 for getting an odd number or 6 on
    each roll, and loses $1 otherwise? What is the probability of winning in
    these games if the gambler starts with $1?

8   In 1981, Kathleen Morris and Harold Lecar presented a simple model to
    generate action potentials of the membrane established in a neuron [12].

This model consists of three channels: potassium, leak, and calcium, which are modelled by:

$$C_M \frac{dV}{dt} = I_{app} - g_L(V - E_L) - g_K n(V - E_K)$$
$$- g_{Ca} m_\infty(V)(V - E_{Ca}),$$
$$\frac{dn}{dt} = \Phi(n_\infty(V) - n)/\tau_n(V),$$

where

$$m_\infty(V) = \frac{1}{2}\left(1 + \tanh\left(\frac{V - V_1}{V_2}\right)\right),$$
$$\tau_n(V) = \frac{1}{\cosh\left(\frac{V - V_3}{2V_4}\right)},$$
$$n_\infty(V) = \frac{1}{2}\left(1 + \tanh\left(\frac{V - V_3}{V_4}\right)\right).$$

In these equations, V denotes the membrane potential and n is used for the probability that the potassium channel (K) is conducting. The parameters I_{app} and C_M are applied current and membrane capacitance, respectively; g_L, g_K, and g_{Ca} are conductances through membrane channels for leak, potassium, and calcium, respectively; E_L, E_K, and E_{Ca} represent the equilibrium potential of leak, potassium, and calcium ion channels, receptively; V_1, V_2, V_3, and V_4 are parameters specific to voltage-clamp data; and Φ is called a reference frequency. The values of these parameters are shown in Table 9.1.

The equivalent circuit for this model is shown in Figure 9.8. I_{app} can be employed as a bifurcation parameter to observe Hopf bifurcation. In Figure 9.9 we have plotted several trajectories to observe limit cycles around the critical point for different values of I_{app} using the parameter values in Table 9.1. Employ the new parameter values $V_3 = 12$, $V_4 = 17.4$, and $\Phi = 0.23$ to develop a discrete method to simulate and plot these curves for I_{app} in the set $[0, 40, 41, 100]$. In this case, we observe that a homoclinic orbit emerges around $I_{app} = 41$.

9   The following model describes the interaction between bacteria, activated neutrophils, and cytokines [47]:

$$\frac{dP}{dt} = r_p P\left(1 - \frac{P}{P_\infty}\right) - r_{cl}\frac{P^n}{(P^n + k_c^n)}N\,P - \delta_p P,$$
$$\frac{dN}{dt} = r_n N_R P(T) - \delta_N N,$$
$$\frac{dT}{dt} = \frac{K_{max}N}{(C_N + N)} - \delta_T T,$$

Table 9.1 Parameter values for the Morris–Lecar model.

Parameter	Hopf	Homoclinic	Unit
C_M	20	20	$\mu F\ s/cm^2$
g_{Ca}	4.4	4.4	$\mu F/cm^2$
g_K	8	8	$\mu F/cm^2$
g_L	2	2	$\mu F/cm^2$
V_1	−1.2	−1.2	mV
V_2	18	18	mV
V_3	2	12	mV
V_4	30	17.4	mV
E_{Ca}	120	120	mV
E_K	−84	−84	mV
E_L	−60	−60	mV
Φ	0.04	0.23	–

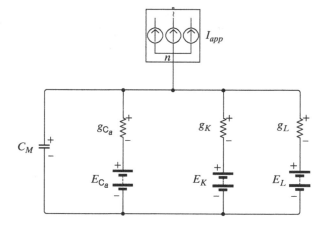

Figure 9.8 Equivalent circuit for the Morris–Lecar model.

where P, N, and T represent the number of bacteria, number of activated neutrophils, and TNF-alpha concentration, respectively. The parameter r_p represents the bacteria growth rate, P_∞ the bacteria carrying capacity, r_{cl} the rate at which neutrophils kill bacteria, n the Hill coefficient, k_{cl} the extent of bacteria binding to neutrophils, δ_p the decay rate of bacteria due to other factors, r_n the recruitment rate of activated neutrophils

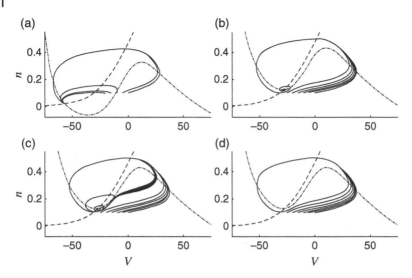

Figure 9.9 Hopf trajectories of the Morris–Lecar model with (a) $I_{app} = 10$; (b) $I_{app} = 87$; (c) $I_{app} = 88.309$; (d) $I_{app} = 90$.

by released cytokine and existing bacteria, N_R the number of resting neutrophils in the system, δ_N the natural death rate of activated neutrophils, K_{max} the maximum release rate of cytokine (TNF-alpha), C_N the neutrophil concentration (counts) at which the release rate of TNF-alpha is half of its maximum, and δ_T the natural decay rate of TNF-alpha. Find the critical points of this system and analyze their stability with $n = 0, 1, 2$.

References

1 Murray E Alexander and Seyed M Moghadas. $O(\ell)$ shift in Hopf bifurcations for a class of non-standard numerical schemes. *Electronic Journal of Differential Equations*, 12:9–19, 2005.

2 Murray E Alexander, Arthur R Summers, and Seyed M Moghadas. Neimark–Sacker bifurcations in a non-standard numerical scheme for a class of positivity-preserving ODEs. *Proceedings of the Royal Society of London A: Mathematical, Physical and Engineering Sciences*, 462(2074):3167–3184, 2006.

3 Linda JS Allen. *An Introduction to Stochastic Processes with Applications to Biology*. CRC Press, Boca Raton, FL, 2010.

4 Linda JS Allen, Bernd Aulbach, and Elaydi Saber. *Difference Equations and Discrete Dynamical Systems: Proceedings of the 9th International Conference*. World Scientific, Singapore, 2005.

5 John Joseph Anagnost and Charles A Desoer. An elementary proof of the Routh-Hurwitz stability criterion. *Circuits, Systems, and Signal Processing*, 10(1):101–114, 1991.

6 Roy M Anderson, Robert M May, and B Anderson. *Infectious Diseases of Humans: Dynamics and Control*. Oxford University Press, Oxford, 1992.

7 William E Boyce and Richard C Diprima. *Elementary Differential Equations with Mathematica*. John Wiley & Sons, Inc., 1997.

8 Fred Brauer and Carlos Castillo-Chavez. *Mathematical Models in Population Biology and Epidemiology*. Springer, New York, 2001.

9 Lauren B Carrington and Cameron P Simmons. Human to mosquito transmission of dengue viruses. *Frontiers in Immunology*, 5, 2014.

10 L. Devroye. *Non-Uniform Random Variate Generation*. Springer, New York, New York, 1986.

11 David JD Earn, Pejman Rohani, Benjamin M Bolker, and Bryan T Grenfell. A simple model for complex dynamical transitions in epidemics. *Science*, 287(5453):667–670, 2000.

12 G Bard Ermentrout and David H Terman. *Mathematical Foundations of Neuroscience*. Springer, New York, 2010.

Mathematical Modelling: A Graduate Textbook, First Edition. Seyed M. Moghadas and Majid Jaberi-Douraki.
© 2019 John Wiley & Sons, Inc. Published 2019 by John Wiley & Sons, Inc.
Companion Website: www.wiley.com/go/Moghadas/Mathematicalmodelling

13 Richard FitzHugh. Impulses and physiological states in theoretical models of nerve membrane. *Biophysical Journal*, 1(6):445–466, 1961.

14 H Scott Fogler. *Essentials of Chemical Reaction Engineering*. Prentice Hall, Upper Saddle River, NJ, 2010.

15 HI Freedman. *Deterministic Mathematical Models in Population Ecology*. Marcel Dekker, New York, 1980.

16 Anne Greenbaum and Timothy P Chartier. *Numerical Methods: Design, Analysis, and Computer Implementation of Algorithms*. Princeton University Press, Princeton, NJ, 2012.

17 Duane J Gubler and Gary G Clark. Dengue/dengue hemorrhagic fever: the emergence of a global health problem. *Emerging Infectious Diseases*, 1(2):55, 1995.

18 Richard Haberman. *Mathematical Models: Mechanical Vibrations, Population Dynamics, and Traffic Flow*. Society for Industrial and Applied Mathematics, Philadelphia, 1998.

19 Robert L Harrison, Carlos Granja, and Claude Leroy. Introduction to Monte Carlo simulation. *AIP Conference Proceedings*, 1204:17–21, 2010.

20 Karel Hasík. On a predator–prey system of Gause type. *Journal of Mathematical Biology*, 60(1): 59–74, 2010.

21 Herbert W Hethcote. The mathematics of infectious diseases. *SIAM Review*, 42(4):599–653, 2000.

22 Majid Jaberi-Douraki, Massimo Pietropaolo, and Anmar Khadra. Predictive models of type 1 diabetes progression: understanding T-cell cycles and their implications on autoantibody release. *PloS ONE*, 9(4):e93326, 2014.

23 Patrick MA James, M-J Fortin, BR Sturtevant, A Fall, and D Kneeshaw. Modelling spatial interactions among fire, spruce budworm, and logging in the boreal forest. *Ecosystems*, 14(1):60–75, 2011.

24 Diána Knipl and Seyed M Moghadas. The potential impact of vaccination on the dynamics of dengue infections. *Bulletin of Mathematical Biology*, 77 (12):2212–2230, 2015.

25 Hans Petter Langtangen and Geir K Pedersen. *Scaling of Differential Equations*. Springer, Cham, 2016.

26 Donald Ludwig, Dixon D Jones, and Crawford S Holling. Qualitative analysis of insect outbreak systems: The spruce budworm and forest. *Journal of Animal Ecology*, 47:315–332, 1978.

27 MathWorks. Matlab primer. https://www.mathworks.com/help/pdf_doc/matlab/getstart.pdf, 2017.

28 Deborah G McCullough, Richard A Werner, and David Neumann. Fire and insects in northern and boreal forest ecosystems of North America. *Annual Review of Entomology*, 43(1):107–127, 1998.

29 N Metropolis. The beginning of the Monte Carlo method. *Los Alamos Science*, (Special Issue 15), 1987.

30 Ronald E Mickens. *Nonstandard Finite Difference Models of Differential Equations*. World Scientific, Singapore, 1994.

31 SM Moghadas and ME Alexander. Bifurcation and numerical analysis of a generalized Gause-type predator-prey model. *Dynamics of Continuous, Discrete and Impulsive Systems, Series B*, 13(5):533, 2006.

32 SM Moghadas, ME Alexander, BD Corbett, and AB Gumel. A positivity-preserving Mickens-type discretization of an epidemic model. *Journal of Difference Equations and Applications*, 9(11):1037–1051, 2003.

33 SM Moghadas, ME Alexander, and BD Corbett. A non-standard numerical scheme for a generalized Gause-type predator–prey model. *Physica D: Nonlinear Phenomena*, 188(1):134–151, 2004.

34 Jacques Monod. *Recherches sur la croissance des cultures bacteriennes*. Hermann, Paris, 1942.

35 Jinichi Nagumo, Suguru Arimoto, and Shuji Yoshizawa. An active pulse transmission line simulating nerve axon. *Proceedings of the IRE*, 50(10):2061–2070, 1962.

36 David L Nelson, Albert L Lehninger, and Michael M Cox. *Lehninger Principles of Biochemistry*. W.H. Freeman, New York, 2008.

37 Aaron Novick and Leo Szilard. Description of the chemostat. *Science*, 112(2920):715–716, 1950.

38 David W Oxtoby, H Pat Gillis, and Laurie J Butler. *Principles of Modern Chemistry*. Cengage Learning, 2015.

39 Lawrence Perko. *Differential Equations and Dynamical Systems*. Springer Science & Business Media, New York, 2013.

40 Nicholas G Reich, Sourya Shrestha, Aaron A King, Pejman Rohani, Justin Lessler, Siripen Kalayanarooj, In-Kyu Yoon, Robert V Gibbons, Donald S Burke, and Derek AT Cummings. Interactions between serotypes of dengue highlight epidemiological impact of cross-immunity. *Journal of the Royal Society Interface*, 10(86):20130414, 2013.

41 FJ Richards. A flexible growth function for empirical use. *Journal of Experimental Botany*, 10(2):290–301, 1959.

42 Raina Robeva and David Murrugarra. The spruce budworm and forest: a qualitative comparison of ODE and Boolean models. *Letters in Biomathematics*, 3(1):75–92, 2016.

43 Sheldon M Ross. *Introduction to probability models*. Academic Press, 2014.

44 Edward John Routh. *A Treatise on the Stability of a Given State of Motion: Particularly Steady Motion*. Macmillan and Company, London, 1877.

45 Rüdiger Seydel. *Practical Bifurcation and Stability Analysis*. Springer, New York, 2010.

46 Zhenzhen Shi, Chih-Hang J Wu, David Ben-Arieh, and Steven Q Simpson. Mathematical model of innate and adaptive immunity of sepsis: A modeling and simulation study of infectious disease. *BioMed Research International*, 2015, 2015.

47 Zhenzhen Shi, Stephen K Chapes, David Ben-Arieh, and Chih-Hang Wu. An agent-based model of a hepatic inflammatory response to salmonella: A computational study under a large set of experimental data. *PloS ONE*, 11(8):e0161131, 2016.

48 Gordon D Smith. *Numerical Solution of Partial Differential Equations: Finite Difference Methods.* Oxford University Press, 1985.

49 Gilbert Strang. *Introduction to Linear Algebra.* Wellesley-Cambridge Press, Wellesley, MA, 1993.

50 Andrew Stuart and Anthony R Humphries. *Dynamical Systems and Numerical Analysis.* Cambridge University Press, Cambridge, 1998.

51 Aslak Tveito and Ragnar Winther. *Introduction to Partial Differential Equations: A Computational Approach.* Springer, Berlin, 2004.

52 John J Tyson. Some further studies of nonlinear oscillations in chemical systems. *Journal of Chemical Physics*, 58(9):3919–3930, 1973.

53 Pierre-François Verhulst. Notice sur la loi que la population suit dans son accroissement. *Correspondance Mathématique et Physique*, 10:113–121, 1838.

54 Annelies Wilder-Smith, Karl-Erik Renhorn, Hasitha Tissera, Sazaly Abu Bakar, Luke Alphey, Pattamaporn Kittayapong, Steve Lindsay, James Logan, Christoph Hatz, Paul Reiter, et al. Denguetools: innovative tools and strategies for the surveillance and control of dengue. *Global Health Action*, 5(1):17273, 2012.

55 Darren J Wilkinson. *Stochastic Modelling for Systems Biology.* CRC Press, Boca Raton, FL, 2011.

56 Jacques Leopold Willems. *Stability Theory of Dynamical Systems.* Nelson, London, 1970.

Index

Mathematical Modelling: A Graduate Textbook, First Edition. Seyed M. Moghadas and Majid Jaberi-Douraki.
© 2019 John Wiley & Sons, Inc. Published 2019 by John Wiley & Sons, Inc.
Companion Website: www.wiley.com/go/Moghadas/Mathematicalmodelling